OXFORD MONOGRAPHS ON GEOLOGY AND GEOPHYSICS

Chemical Change in Deforming Materials

BRIAN BAYLY

Department of Geology
Rensselaer Polytechnic Institute

New York   Oxford
OXFORD UNIVERSITY PRESS
1992

Oxford University Press

Oxford   New York   Toronto
Delhi   Bombay   Calcutta   Madras   Karachi
Kuala Lumpur   Singapore   Hong Kong   Tokyo
Nairobi   Dar es Salaam   Cape Town
Melbourne   Auckland   Madrid

and associated companies in
Berlin   Ibadan

Published by Oxford University Press, Inc.
200 Madison Avenue, New York, New York 10016

Oxford is a registered trademark of Oxford University Press

Library of Congress Cataloging-in-Publication Data
Bayly, M. Brian, 1929–
Chemical change in deforming materials / Brian Bayly.
p.   cm. — (Oxford monographs on geology and geophysics ; no. 21)
Includes bibliographical references and index.
ISBN 0-19-506764-9
1. Materials—Compression testing.   2. Deformations (Mechanics)
3. Chemical equilibrium.   4. Geochemistry.   I. Title.   II. Series.
TA417.7.C65B39 1992   620.1'1242—dc20   92-3822

2 4 6 8 9 7 5 3 1

Printed in the United States of America
on acid-free paper

To Hans

PREFACE

Several years ago I found myself puzzled, and this book is an outcome of that puzzlement. The core of the puzzle lies, as does so much else, in the work of Willard Gibbs.

Gibbs considered a cube of a solid material; each pair of opposite faces was in contact with a fluid and the three fluids were at three different pressures. Gibbs proposed (1878, eqns. 393–395) that for equilibrium to exist, it was necessary for the chemical potential of the material of the solid in one fluid to be different from the chemical potential of the material of the solid in the second fluid. Saying nothing about the chemical potential of the material of the solid in the solid, he left a predicament that, for the science community as a whole, is still unresolved today. At equilibrium, one would expect the potential of the material of the solid to be the same in the solid as in any adjacent fluid, but how can it be the same as in three fluids at once? It was inability to walk away from that question that prompted the thoughts in the following chapters. Of course, many of the thoughts were picked up from other people and in this connection I should like to comment briefly first on policy, then on history.

The policy adopted has two threads: first, no attempt is made at a comprehensive review of what has been written on the topic by other people; second, in the main text, references to other writing have been kept to a minimum. Both of these derive from the wish to present some fundamental ideas in a plain and simple form. There is also the possibility or suggestion that, for all its high quality, earlier writing does not begin at the beginning. (Digressing briefly, I suggest that one needs to begin by taking a material component's potential as direction-dependent, thus being multivalued rather than single-valued at a point, and by emphasizing nonrecoverable deformation, where energy is dissipated, to the exclusion of recoverable or elastic deformation, where energy is stored. A runner in a short race uses two starting-blocks; if these two are indeed the two starting-blocks for the topic in hand, I know of no treatment that has got off to a good start by using both. Of course, my knowledge of what has been written is fragmentary and I apologize to any person who deserves mention at this point whom I am passing over.)

As regards history, let us look at the use of multi-valued potentials (I know too little to survey the extensive literature in which a component's potential is taken as single-valued). Key papers are two by Hans Ramberg: "The Gibbs' free energy of crystals under anisotropic stress, a possible cause for preferred mineral orientation," *Anais da Escola de Minas de Ouro Preto, 32* (1959) 1–12; and "Chemical thermodynamics in mineral studies," *Physics and Chemistry of the Earth, 5* (1963) 225–252; and two by Ray M. Bowen: "Toward a thermodynamics and mechanics of mixtures," *Archive for Rational Mechanics and Analysis, 24* (1967) 370–403; and "Theory of mixtures," in *Continuum Physics,* vol. 3, ed. A. C. Eringen, New York, Academic Press, (1976) pp. 1–127. In each pair, the second is easier to get hold of: in Ramberg's case, a more widely distributed journal and in Bowen's case a more complete and self-contained presentation. Ramberg introduces the idea of a component's potential at a point being multivalued; he mentions specifically the three principal values. Bowen introduces the idea of a component's potential being a tensor (with principal values matching those proposed by Ramberg).

The other starting-block is emphasis on deformation processes that are dissipative rather than elastic. Links between a material's chemistry and elastic deformation have been widely sought, but the simpler links between change of chemistry and creep or dissipative or viscous deformation have been treated only more recently by G. B. Stephenson ("Deformation during interdiffusion," *Acta Metallurgica et Materialia, 36* (1988) 2663–2683). This book is an attempt to combine the ideas of Ramberg with those of Stephenson.

Two more policies should be mentioned: one concerns language and the other individuality. Regarding language, we note the difference between legal and plain language. Each has virtues, each is needed in its proper place. When A wishes to introduce a new concept to B, to begin with plain language is often efficient. It may be necessary later to switch to more sophisticated language to increase the precision with which the concept is expressed, but to *start* by means of statements expressed with a high level of exactness (and with the sophisticated phrasing that that requires) is not necessarily efficient. Practice of course varies with the group addressed and the speaker's or writer's intention. Here an effort is made to express the ideas in plain language, with diagrams and analogies. A reader may benefit more from a complete set of ideas roughly sketched than from an incomplete set of ideas more precisely expressed.

So we come to the last prefatory remark, which returns us to the opening. What follows is not authoritative; the intention is to lay some ideas before the reader that he or she may cogitate about independently.

He that made us ... gave us not ... reason to fust in us unused.

SHAKESPEARE, Hamlet IV, 4

The authority of a thousand is not worth the humble reasoning of a single individual.

GALILEO, Letters on sunspots (Rome, Lincean Academy 1613)

It is important that students bring a certain ragamuffin, barefoot irreverence to their studies; they are not here to worship what is known but to question it.

BRONOWSKI, Ascent of Man (London, BBC 1975)

At some points in the text, wording is chosen for terseness and at some points for clarity; in both instances the effect may be to imply definiteness or to seem didactic, but this is not the intention. I have tried to put the ideas in a form that the reader can *grasp*; it is the reader's business to decide how much or how little to *accept*. I would be happy if the reader would treat what follows as raw clay, and use it to mold something more durable or substantial for himself or herself.

Acknowledgments

It is a pleasure to record the part played by other people, both in general encouragement and in more technical discussions.

In the first group I wish to recognize my parents, my wife, Geoffrey Crowson and the Provost and Fellows of King's College, Cambridge. The value of individual people and the value of having people who are different from each other: in upholding these, the people named directed my course in life.

Colleagues who helped at various times with patience and good humor are Eric Wild, Don Drew, Ray Fletcher, Florian Lehner and George Fisher; many students, of whom I would like to name Sharon Finley, Frank Florence, Thomas Menard and Ray Donelick; and also Joseph Kestin, Jean Cogné, Peter Cobbold, Enrique Merino, Bob Wintsch and Graham Borradaile.

In both roles Hans Ramberg stands alone. He has enriched the sciences for all of us and, by a happy chance, plays a role in my life for which I never cease being glad. We are all in his debt.

Rensselaer Polytechnic Institute B.B.
May 1992

CONTENTS

II SIMULTANEOUS DEFORMATION AND DIFFUSION

III APPLICATION: MOVEMENTS ALONG ONE DIRECTION

IV EXTENSIONS

18. Cylindrical inclusions, 187

19. Review of strategies, 199

20. Further extensions, 208

SYMBOLS

A	(i) an area
	(ii) a factor in an exponential term Ae^{kx}, etc., especially, the governing stress magnitude in a stress that varies exponentially
A	an atomic species in a binary material (A, B) or a compound (A, B)X
A_i	activity coefficient of component i
a	volume fraction of component A in a material
a^a	volume fraction of an imaginary additive
B	the length-scale factor in an exponential term $Ae^{x/B}$
B	an atomic species accompanying species A (q.v.)
b	volume fraction of component B in a material
C^i, C_i	concentration of species i = (number of atoms of i in sample)/ (total number of atoms in sample). In a compound (A, B)X,

$$C^A = \frac{\text{number of atoms of A}}{\text{total number of A + B}}$$

C_0	a concentration magnitude that is uniform through space
c	the extreme value of a fluctuation in composition from point to point
D	a diffusion coefficient; see Note following this list
D^i	a diffusion coefficient; see Note following this list
E	an elastic property
E^*, E^a	a coefficient for interdiffusion of two atomic species
E_I, E_{II}, E_{III}	strain rate invariants
e	linear strain rate
e_1, e_2, e_3	principal strain rates at a point
e_0	(i) mean strain rate at a point, $e_0 = (e_1 + e_2 + e_3)/3$
	(ii) a strain rate that is arbitrarily fixed and is uniform through space (Chapter 13 on)
e_{ii}	linear strain rate along direction i (no summation convention)
e_{ij}	shear strain rate on a plane normal to i along direction j ($i \neq j$)
e_n	linear strain rate along direction $\mathbf{n}$
e_n'	the difference $e_n - e_0$, $e_n - $ (mean strain rate)
F	force
G	(i) free enthalpy or Gibbs free energy
	(ii) elastic shear modulus $\sigma_n'/2\varepsilon_n'$

g, h   two coexisting phases

H the governing stress magnitude in a term He^{kx} that describes an
              exponentially varying stress

j (i) amplitude of a fluctuation in pressure (Chapter 8)
              (ii) wavenumber as in $C = C_0 + c \sin jx$

K a diffusion coefficient: $\partial(\text{vol})/(\text{vol})\,\partial t = K\,\nabla^2 P$. See Note follow-
              ing this list

K^a coefficient for interdiffusion of two components

K^i diffusion coefficient for component i in a composite material

$K*$ joint diffusion coefficient for two or more species moving together

K_1 diffusion coefficient: flux $= K_1\,\partial(\text{conc})/\partial x$. See Note following
              this list

K_2 diffusion coefficient: particle velocity $= K_2\,\partial\mu/\partial x$. See Note fol-
              lowing this list

K elastic bulk modulus $\sigma_0/3\varepsilon_0$

k (i) a general-purpose proportionality constant
              (ii) a diffusion coefficient: $\partial(\text{vol})/(\text{vol})\,\partial t = k\,\nabla^2\mu$. See Note fol-
              lowing this list
              (iii) inverse length in a term e^{kx}
              (iv) a factor relating viscosity to composition

L the characteristic length of a material $2(NK)^{1/2}$

L_i the characteristic length of material i in a composite material

L_m a mean value formed from two lengths of type L_i

L_0 the length of a circular arc that is a quarter-circle

l a small length

l, m, n a set of orthogonal directions

m (i) a mass of material
              (ii) one of the set l, m, n

m_i mass of component i

N viscosity $\sigma'_n/2e'_n$

N_i the viscosity of part i of a composite material

$N*$ the viscosity of a mixture

n (i) the extreme value of a fluctuation in viscosity from point to point
              (ii) one of the set l, m, n

$\mathbf{n}$ a direction or unit vector

n_1, n_2, n_3 components of $\mathbf{n}$ along directions 1, 2, and 3

P pressure

P_0 a uniform reference pressure

P_e one of the pair (T_e, P_e) at which two phases are in equilibrium

P_i a pressure imposed on a sample formerly in equilibrium at P_e

Q quantity, for example quantity of heat or mass

R (i) the gas constant
              (ii) in Chapters 12 and 13, the radius of a quarter-circle with
              arc-length $L_0, = 2(NK)^{1/2}$. From Chapter 14 onward, L is
              used in place of R for this quantity

R (*cont.*)	(iii) a stress difference $\frac{1}{2}(\sigma_{zz} + \sigma_{yy}) - \sigma_{xx}$ (Appendix 13A only)
R	a diffusing component (Chapter 4 only)
S	(i) entropy
	(ii) the stress difference $\sigma_{zz} - \sigma_{xx}$
S_0	a stress magnitude that is uniform through space
S_I, S_{II}, S_{III}	stress invariants
s	the extreme value of a fluctuation in stress from point to point
s'	$= s(1 + j^2 R^2/2)$
T	(i) temperature
	(ii) a stress difference $\frac{1}{2}(\sigma_{zz} - \sigma_{yy})$ (Appendix 13A only)
T_0	a uniform reference temperature
T_e	one of the pair (T_e, P_e) at which two phases are in equilibrium
t	time; δt, a short period of time
U	internal energy
V	(i) volume of a specified sample
	(ii) volume of 1 kg or 1 kg-mole of material, according to context
V_i	volume of 1 kg or 1 kg-mole of species i, according to context
X^i, X_i	concentration of component i, same as C_i
X_I, X_{II}	concentration invariants
x	volume-fraction of component X in a material
x, y, z	coordinates of a point
x_1, x_2, x_3	a set of orthogonal axes
α	an angle in the xz plane or x_1x_3 plane
α_1, α_2	angles measured away from a line in two orthogonal planes
α, β, γ	angles between a line and three axes x, y, z or x_1, x_2, x_3
α_i	activity of component i, $= A_i X_i$
ε	linear strain, $=$ (change in length)/(original length)
$\varepsilon_1, \varepsilon_2, \varepsilon_3$	principal strains at a point
ε_0	mean strain at a point, $\varepsilon_0 = (\varepsilon_1 + \varepsilon_2 + \varepsilon_3)/3$
ε_{ii}	linear strain along direction i (no summation convention)
ε_{ij}	shear strain on a plane normal to i along direction j $(i \neq j)$
ε_n	linear strain along direction n
ε_n'	the difference $\varepsilon_n - \varepsilon_0$, $\varepsilon_n -$ mean strain
θ	a general-purpose angle
θ_k	an angle in the plane containing directions i and j
θ_1, θ_2	angles measured away from a line in two orthogonal planes
λ	wavelength of a sine-wave
μ	chemical potential
μ_i, μ_{ii}	chemical potential associated with direction i (no summation convention)
μ_n	chemical potential associated with direction n
μ_1, μ_2, μ_3	principal values of μ at a point
μ_0	(i) chemical potential in a reference state
	(ii) mean value, $\mu_0 = (\mu_1 + \mu_2 + \mu_2)/3$

ρ_i	number of kg-moles of species i in $1\,\mathrm{m}^3$ of material
σ	a stress component
$\sigma_1, \sigma_2, \sigma_3$	principal stresses at a point
σ_0	mean stress at a point, $\sigma_0 = (\sigma_1 + \sigma_2 + \sigma_3)/3$
σ_i, σ_{ii}	normal stress component on a plane normal to i (no summation convention)
σ_{ij}	stress component on a plane normal to i and acting along direction j $(i \neq j)$
σ_{n}	normal stress component on a plane normal to $\mathbf{n}$
σ_{n}'	the difference $\sigma_{\mathrm{n}} - \sigma_0$
σ_{t}	a shear stress component

Note on diffusion coefficients used

Material motion can be driven by a gradient in concentration or pressure or chemical potential. The coefficients mainly used for the three circumstances are D^i, K, and D, as follows. (For K_1, K_2, and k, see below.)

D^i

$$\text{Flux} = D^i \frac{\partial}{\partial x}\left(\text{conc}\ \frac{\text{mol}}{\text{m}^3}\right) \qquad\qquad \frac{\partial Q}{\partial t} = D^i \frac{\partial^2}{\partial x^2}\left(\text{conc}\ \frac{\text{mol}}{\text{m}^3}\right)$$

$$\frac{\text{mol}}{\text{m}^2\text{-sec}} \quad \frac{\text{m}^2}{\text{sec}} \quad \frac{\text{mol}}{\text{m}^4} \qquad\qquad \frac{\text{mol}}{\text{m}^3\text{-sec}} \quad \frac{\text{m}^2}{\text{sec}} \quad \frac{\text{mol}}{\text{m}^5}$$

$$\frac{\partial X}{\partial t} = D^i \frac{\partial^2 X}{\partial x^2} \quad \left[\begin{array}{l} X = \text{mol} \\ \text{fraction} \end{array}\right]$$

$$\frac{1}{\text{sec}} \quad \frac{\text{m}^2}{\text{sec}}\frac{1}{\text{m}^2}$$

K

$$\text{Flux} = K \frac{\partial}{\partial x}(\text{pressure}) \qquad\qquad \frac{\partial Q}{\partial t} = K \frac{\partial^2}{\partial x^2}(\text{pressure})$$

$$\frac{\text{m}^3}{\text{m}^2\text{-sec}} \quad \frac{\text{m}^2}{\text{Pa-sec}} \quad \frac{\text{Pa}}{\text{m}} \qquad\qquad \frac{\text{m}^3}{\text{m}^3\text{-sec}} \quad \frac{\text{m}^2}{\text{Pa-sec}} \quad \frac{\text{Pa}}{\text{m}^2}$$

$$= \frac{\text{volume}}{\text{strain rate}} \quad \frac{\partial(\text{vol})}{(\text{vol})\,\partial t}$$

$$D \qquad\qquad\qquad e = \frac{\partial l}{l\,\partial t} = \quad D \quad \frac{\partial^2}{\partial x^2}\ \text{(potential)}$$

$$\frac{1}{\text{sec}} \qquad \frac{\text{mol}}{\text{Pa-m-sec}} \qquad \frac{J}{\text{mol-m}^2}$$

$$= \text{linear strain rate}$$

For component A, $K^A = 3D^A V^A = aD^{iA} V^A/RT$, where a = volume fraction of component A.

K_1 and K_2 are used briefly in Chapter 3. K_1 is the same as D^i; $K_2 = D^i/RT = K^A/aV^A = 3D^A/a$.

k is used briefly in Chapter 10, $= 3D = K/V$; volume strain rate $\partial(\text{vol})/(\text{vol})\,\partial t = k\,\partial^2\mu/\partial x^2$.

Basis for the relation $K^A = 3D^A V^A$. $\partial\mu^A/\partial P = V^A$; also in an isotropic process, volume strain rate $= 3 \times$ linear strain rate.

Basis for the relation $K^A = aD^{iA} V^A/RT$. For pressure-driven motion, influx of A in $(\text{m}^3/\text{m}^3\text{-sec})$

$$= K^A\,\partial^2 P/\partial x^2 = (K^A/V^A)\frac{\partial^2\mu^A}{\partial x^2} \tag{i}$$

Also, for composition-driven motion, influx of A in $(\text{m}^3/\text{m}^3\text{-sec})$

$$= (\text{influx in mol/m}^3\text{-sec})V^A = D^{iA} V^A \frac{\partial^2(\text{conc mol/m}^3)}{\partial x^2} \tag{ii}$$

To compare (i) and (ii) we note that if conditions permit the term $(\partial X^A/X^A\,\partial x)^2$ to be neglected,

$$\frac{\partial^2\mu^A}{\partial x^2} = \frac{RT}{X^A}\frac{\partial^2 X^A}{\partial x^2}$$

$$= \frac{RT}{(\text{conc mol/m}^3)}\frac{\partial^2}{\partial x^2}(\text{conc mol/m}^3)$$

$$= \frac{RTV^A}{a}\frac{\partial^2}{\partial x^2}(\text{conc mol/m}^3)$$

Hence for composition-driven motion, influx of A in $(\text{m}^3/\text{m}^3\text{-sec})$

$$= \frac{aD^{iA}}{RT}\frac{\partial^2\mu^A}{\partial x^2} \tag{iii}$$

For equivalence of (i) and (ii), $K^A = aD^{iA} V^A/RT$.

CHEMICAL CHANGE IN DEFORMING MATERIALS

1

OVERVIEW AND PREVIEW
OF CONCLUSIONS

The purpose of this book is to fill something of a gap. In general, thermodynamics has been a great success and has provided a means of understanding and predicting material behavior of almost all kinds at the macroscopic level. Even when thermodynamic statements were limited to equilibrium states they were widely useful, and with extension to nonequilibrium states almost all behaviors that a person might observe directly became accessible to theory. But there has been and is one resistive point: if a cylinder of material is more strongly compressed along its length than radially, it is in a nonequilibrium state no matter how ideal its condition in other respects, and the effect of this type of nonequilibrium has not been successfully explored.

The physical consequences are, of course, well known; the cylinder deforms in ways successfully described in almost all respects by the methods of continuum mechanics. But the chemical consequences are less well known. For example, suppose the cylinder contains iron and is surrounded by some second iron-bearing phase; suppose further that before the cylinder is compressed axially, the cylinder and its surroundings are in equilibrium. When the axial compression is imposed, how is the equilibrium disturbed and what processes begin to run? The purpose of the book is to provide the outline of a comprehensive approach to this question.

The question has been discussed extensively in technical journals and in complicated ways. The stimulus for this book is the belief that the topic need not be so complicated. There are two equations that describe the stresses in the cylinder that have up to now not been used; using these neglected equations provides a point of view not taken by other writers, and it is the fresh point of view that permits certain simplicities to be seen (the key equations are 6.3 and 8.10). Of course, we make headway only to a limited extent; not all problems are answered, not all complications are resolved. The existence of a central and unresolvable complication is recognized toward the end of this overview, in the section on Continuum Behavior and Atoms.

Materials considered. We shall give most of our attention to materials that are in the process of deforming. The material one imagines depends on the time scale envisaged: the ice of a glacier deforms appreciably in a year, whereas silicone putty deforms appreciably in a few minutes; water deforms even more readily but we shall not consider anything as mobile as water because here forces related to the material's acceleration can be as large as the other forces present. We shall stay within circumstances where the material's acceleration can be neglected. Hot ceramic materials are considered, both crystalline and glassy; also the naturally occurring silicates and oxides known to geologists; also polymers, metals and semiconductors. We make gross simplifying assumptions so that the differences among these materials are submerged; the material treated in detail is in fact an imaginary material simpler than any of these. But the materials listed are the target at which the simplified discussions are aimed.

Elastic behavior. At least for a start, we shall also choose circumstances where elastic behavior can be neglected. A stressed elastic material differs from the same material when unstressed in respect of bond lengths and quantity of energy stored per gram or kilogram; but we shall assume that these changes hardly affect the way the material flows or creeps—at least for a start. After discussion of the flow behavior, the question of elastic behavior will be taken up.

Chemistry. We shall assume rather simple chemistry. In some of the discussion, the material will be as simple as ice, that is to say of a single, fixed composition; and at other times we shall consider binary mixtures such as the sulfide $(Zn, Fe)S$—which can be thought of as a mixture of ZnS and FeS—or a polymer mixture of trifluorethylene and tetrafluorethylene. With such a binary material, there is the possibility of variation in composition within a sample. One might say that the purpose of the book is to learn to understand how nonhydrostatic stress can affect such variations in composition.

Stresses. Inside a deforming continuum (glacier or putty lump, as before), if one could inspect a small area, one would find the material on one side of the area pressing and dragging on the material on the other side: normal to the element of area considered, there is either a compression or a tension, and in addition there is a stress tangential to the area considered. We shall give attention mainly to the normal stress and mainly to compressions (less emphasis on tensions). Among the circumstances to be considered is the special case of hydrostatic pressure, where, whatever plane one considers through a single point inside the continuum, one finds the same compressive stress across it; and also, most importantly, the more general case where, at a single point, some planes are at a higher level of compressive stress and some at a lower level.

Chemical potential. In classical thermodynamics, equilibrium states are emphasized. In a state of equilibrium, each component of a material has a chemical potential that takes the same single value at every point in the system considered. A component's chemical potential depends on the pressure and temperature and on the component's concentration. For example, in the sulfide already mentioned, (Zn, Fe)S, the component ZnS would have a chemical potential; and we can imagine two separate samples, each in equilibrium within itself, such that in the sample with higher concentration of ZnS the potential of the ZnS component was higher, and in the sample with lower concentration it was lower.

Similarly with pressure: if two samples had the same concentration of ZnS but one sample were at a higher pressure, the potential of ZnS would be higher in that sample than in the other.

Preview of conclusions

A preliminary view of the conclusions reached in later chapters is given in the next four sections.

Deformation

The concept of a *wafer* is important. The behavior of a gas can be understood by thinking about single atoms, but the behavior of a glacier cannot. In a glacier or other stiff, deforming material, if two points in the material get closer, then two *planes* in the material have to get closer, because doing continuum mechanics involves assuming that the material is continuous, which involves assuming that *planes* retain their continuity from moment to moment. And if two planes get closer together, one needs to get rid of a whole *wafer* of material from between them. Figure 1.1 shows the role of wafers in allowing a continuous material to shorten along some direction.

If the compression on a small portion inside a continuous material is locally high, wafers can escape (i) to a new *location* by making a short journey through space, or (ii) to a new *orientation* while remaining at the same site; see Figure 1.2. The result of operation (i) is called diffusion or self-diffusion, whereas the result of operation (ii) is called deformation or change of shape "at a point."

A material's stiffness or viscosity is its resistance to change of type (ii), and it must be possible to find a change of type (i) that is exactly as sluggish; that is to say, there must be a distance L_0 such that journeys of type (i)

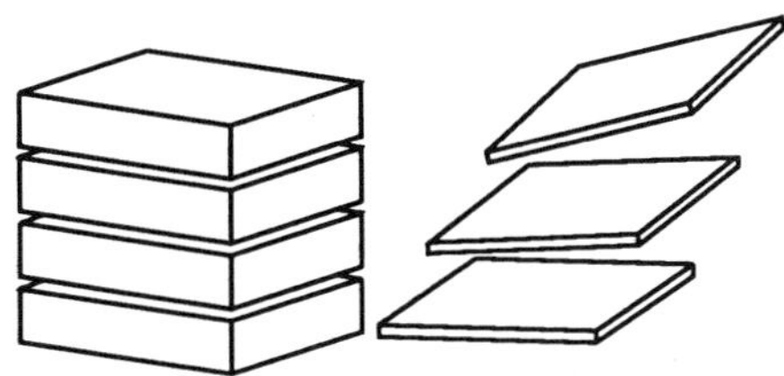

Figure 1.1 Escape of wafers allows a column to shorten vertically. To conceive the deformation of an ideal continuum, one may imagine the scale on which such wafers exist to be infinitesimal.

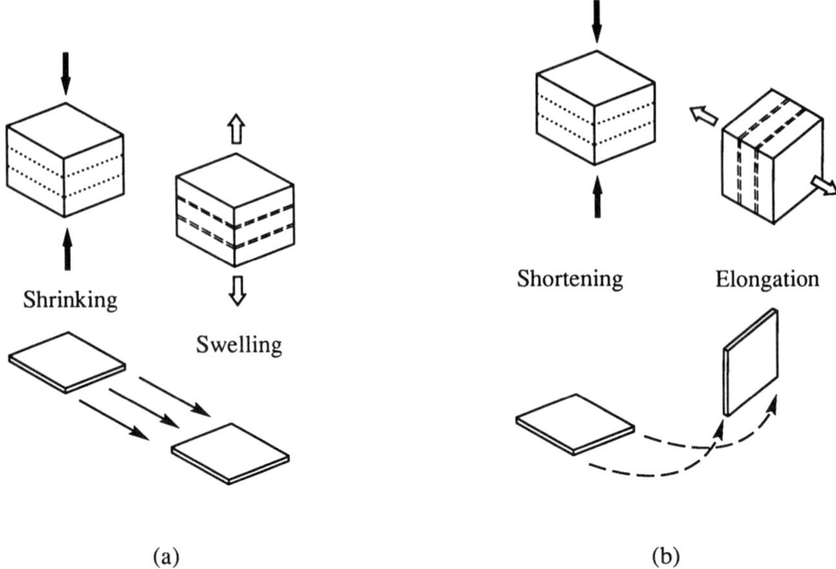

Figure 1.2 Movement of wafers can be imagined as an idealization both in diffusion from one site to another and in change of a sample's shape. In both types of change, the wafers are taken as infinitely small and numerous.

longer than L_0 are more difficult to drive by squeezing than journeys of type (ii), and journeys of type (i) shorter than L_0 are easier to drive than journeys of type (ii).

Such a distance L_0 is an inherent property of the material, along with other properties such as its density and viscosity. For water, L_0 is about 1 nanometer, whereas for fused silica it seems to be 10 or 20 nanometers. Driving material to a new orientation is very like driving it to a new location; whatever the means by which material travels, both operations are described in continuum terms as loss, gain or translation of infinitesimal wafers. Deformation occurs *as if* the transport distance were L_0, while stress-driven self-diffusion operates usually over greater distances.

Chemical potential

Even when a sample is not in equilibrium (and hence some process of adjustment toward equilibrium is running, such as diffusion of material from one point to another) for each component at a point, an *associated equilibrium state* can be identified. Then one can say that the component's behavior will be very much as if it had the chemical potential it *would have* if it were in the associated equilibrium state. For example, material will tend to migrate from a site where the associated equilibrium potential is high to a site where the associated equilibrium potential is lower; this is a reliable prediction even though the associated equilibrium states themselves are wholly imaginary.

For a sample where the stress state is nonhydrostatic, with more compression across some planes at a point than across other planes at the same point, the equilibrium state associated with one plane is different from the equilibrium state associated with another plane. Hence, the chemical potential associated with one plane is different from the chemical potential associated with another; *a range of chemical potentials is associated with a single point*, in the same way as the range of compressive stresses.

The range of chemical potentials, and particularly the difference between the maximum value and the minimum, causes the material to deform. We customarily think of its being the difference in compressive stress that causes the material to deform, but it will be argued that the difference in chemical potential can just as well be treated as the cause. The two differences are proportional, the proportionality factor being just the volume of one mass unit of material.

From the preceding seven paragraphs we learn to see the mechanical deformation of a continuum as being driven by a difference between the chemical potential for one direction and the chemical potential for a second direction. This potential difference makes the material behave as if wafers migrated along short curved paths of an identifiable length L_0.

Interfaces

The simplest physical situation that involves both chemistry and deformation is a cylinder that is being thinned radially and elongated axially in a uniform way by slow viscous deformation. The cylinder is in two parts that meet in the middle at an interface. First we suppose that the two materials are polymorphs, both capable of slow continuous deformation (creep) with different viscosities but the same composition. Then we extend to a situation where the materials are like sphalerite and wurtzite—both having formula (Fe, Zn)S but having different structures and hence different creep viscosities, and also having different Fe/Zn ratios.

We think first of two polymorphs with no chemical variability. If the materials showed no self-diffusion, the profile of radial compressive stress would have a step. But creeping materials do show self-diffusion (because if wafers can behave as in Figure 1.2b they can also behave as in Figure 1.2a). Hence, a real profile is a curve, with no step. The effect of the interface diminishes exponentially away from the interface, and the length scale over which it diminishes to $1/e$ of its maximum value is a multiple of the material's characteristic length L_0.

As Figure 1.3 shows, there is a strong stress gradient near the interface. Beside deforming, material diffuses down this gradient and crosses the interface, or changes phase; in other words, the interface moves through the material. The rate of change of the cylinder's radius and the rate of movement

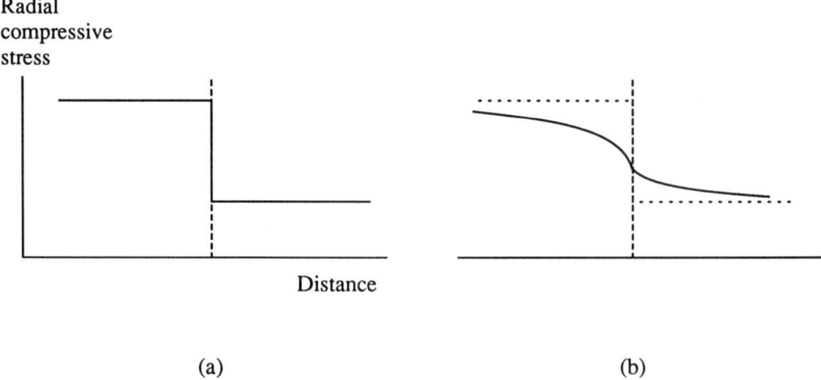

Figure 1.3 Profiles of compressive stress in the neighborhood of an interface: (a) in an imaginary non-diffusing material; (b) in a material that show self-diffusion.

of the interface are linked; if you increase one by squeezing harder, you increase the other by a proportionate amount.

We now think of the cylinder being composed of materials like (Fe, Zn)S. The compositions in the two halves differ and, for the interface to move—for one phase to grow and the other to diminish, there has to be flow of one component to the interface from both sides and flow of the other component away.

A curve of some type such as Figure 1.4c must exist that shows the intensity

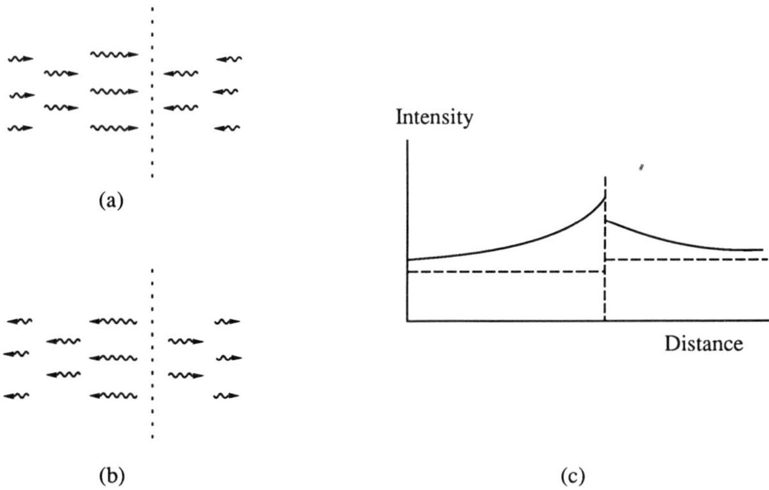

Figure 1.4 Diffusion near an interface. Fluxes of Fe and Zn are necessarily equal and opposite; hence a single pair of curves as in diagram (c) can describe both fluxes.

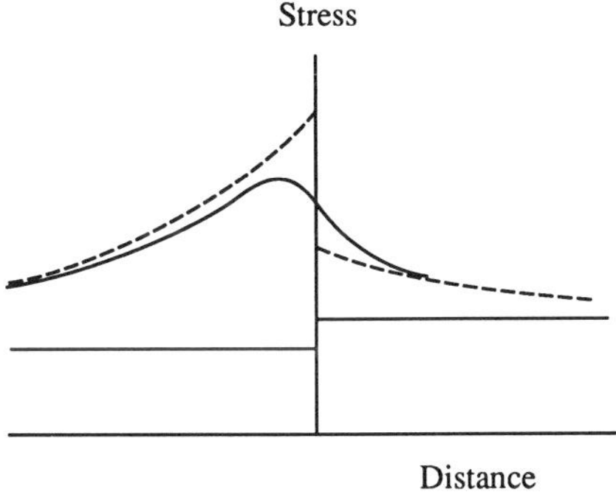

Figure 1.5 Two length scales in the stress profile. The broken lines show a profile that might tend to develop in connection with diffusive exchange of Fe and Zn. But no step at the interface actually develops; the step is rounded off by diffusion of the sulfur host, giving stress gradients that are steep but not infinite.

of the flux of Fe one way and the flux of Zn the other way, the two halves of the curve not necessarily being symmetrical but combining to give the required total effect. The two species Fe and Zn affect the volumes of the unit cells they occupy, so that, if we plotted "rate of change of volume per 100 sulfur atoms," a profile for this quantity would also look like Figure 1.4c, especially if the species that moves towards the interface is the one that occupies a larger volume. Further again, we recall that the cylinder is being constrained: its radius is being diminished slowly in a uniform way. A portion containing, say, 100 sulfur atoms then increases volume only by elongating axially, but in doing this it exerts a radial pressure; a profile of radial compressive stress is a fourth profile that has to be at least approximately of the type in Figure 1.4c. Whenever the interface moves, such a profile must develop.

To understand how the interface behaves, we note that the stress profile cannot be as in Figure 1.3b and as in Figure 1.4c at the same time. The actual stress profile has to be some kind of combination, as in Figure 1.5. It combines a variation on one length scale (as in Figure 1.4c) that is linked to interchange of Fe and Zn *with* a variation on a shorter length scale (as in Figure 1.3b) linked to the deformability of the two ensembles. Each of the latter is probably dominated by the deformability of the swarm of sulfur anions, the sulfur matrix.

Despite the fact that, in general, the stress profile will have two superimposed length scales, there are special conditions in which one or the other effect is suppressed. It is possible to deform a two-part bar of (Fe, Zn)S and to maintain a stress profile of the simple form of Figure 1.3b, or else to deform it, again in specially selected circumstances, so as to maintain a stress profile of the simple form of Figure 1.4c; at least, equations are developed that say so. These two special cases are of no practical value in themselves, but understanding them is a step toward understanding the general case.

Continuum behavior and atoms

Early in this chapter an unresolvable complication was mentioned and Figure 1.5 brings us back to it. The figure shows the superposition of two effects; both diminish away from the interface but they do so on two length scales. The shorter length scale is linked to the material's characteristic length L_0; this is related to the material's deformability, that is, to its behavior as a continuum—stress, strain rate, viscosity, and L_0 itself are all continuum concepts; for understanding processes on this length scale, a mental picture that contains atoms is not necessary, and is in fact an obstacle to a clear grasp of the concepts. By contrast, the longer length scale is linked to the exchange of Fe and Zn; one is forced to imagine atoms, and the energy term $RT \ln X$ is based on atoms. There is an intrinsic clash between the concepts underlying $RT \ln X$ and the concepts underlying L_0.

Despite the above, the stress profile shown in Figure 1.5 is, of course, a single profile; the real-world profile results from behaviors that we understand in part by imagining a continuum and in part by imagining atoms. We can understand the total profile only by using these concepts simultaneously. This book is based on the idea that there is nothing contradictory or self-defeating in doing this. On the other hand, of course, there is merit in keeping the two approaches separate as far as possible and not mixing them until the combination is absolutely needed.

Topics not covered

There are very few descriptions of experimental work and no examples of the application of the ideas to present-day practice. The ideas have in fact not been applied to present-day practice nor even subjected to empirical test. The objective is to sift the ideas and become familiar with them; it is hoped that by the end of the book the reader will be ready to consider the question of how an effective empirical test might be devised. The book is an armchair inquiry; it contains thought-experiments but little reference to actual laboratories.

Another topic not covered is the atomic mechanism by which a material might deform. Vacancy migration, dislocation climb and such topics are purposely avoided as far as possible, being mentioned at just one point for illustrative purposes.

An overview of the overview

We use the idea that at a point in a material, the chemical potential of a component can have a value for one plane through the point that is different from its value for another plane through the same point.

For discussing the material's deformation, we treat it as a continuum, and imagine internal planar elements that remain mathematically continuous. By contrast, for discussing the material's change of chemistry, we treat it as containing countable atoms, some of which exchange places precisely on a one-for-one basis; during an exchange, there is absolutely no continuity between the active atoms and the host or matrix that provides the sites and transport arrangements for the exchange.

At a few points in the development, we suppose that a single material can show both behaviors, continuum and atomic, simultaneously but even here the two behaviors are kept distinct. We specifically avoid trying to link a material's deformation (i.e. its continuum mechanics) to any atomistic mechanism such as vacancy migration by which the deformation might occur.

The next eight chapters assemble the fundamental ideas to be used, which are mostly well established, with just a few novelties. Chapters 10 through 16 contain the developments that result from the point of view taken, and these are amplified and discussed in Chapters 17 through 20.

I
FUNDAMENTALS

2

CHEMICAL POTENTIAL

The purpose of the first chapter was to give an overview of the book's contents: the topics covered, results gained, limitations, and so on. The next seven chapters, starting with this one, give the groundwork on which the main conclusions are based. The intention is to assemble the needed ideas, taking advantage of the fact that an extensive literature exists in which the ideas are established, discussed, restricted, etc. What follows is thus an extract of selected essentials from other documents, rather than being a free-standing and self-contained development of the ideas. The reader is asked to relate the ideas as summarized here to the longer discussions in which they appear elsewhere.

Definition of free energy

The total energy in a portion of material can be split in either of two ways:

Total energy = internal energy + external energy, or

Total energy = free energy + bound energy

In symbols,

$$U + PV = \text{total} = G + TS$$

where U = internal energy of the portion; G = free energy of the portion, specifically the Gibbs free energy or enthalpy; P = pressure; V = volume of the portion; T = temperature; S = entropy of the portion. All the terms except the free energy, G, have independent definitions, so the equations just given define that quantity:

$$G = U + PV - TS \tag{2.1}$$

The equation relates to whatever portion of material one has in view.

Definition of chemical potential

We now suppose that the material has n components and that, in the portion considered, the masses of each are $m_1, m_2, \ldots, m_n$. Then we imagine increasing m_1 by a small amount δm_1 while keeping P, T, and $m_2, m_3, \ldots, m_n$ constant. Let the consequent change in G be δG: then the limit of the ratio $\delta G/\delta m_1$ as $\delta m_1 \to 0$ is the quantity of interest, henceforth written μ_1; it is the

15

chemical potential of component 1 in the material at its current pressure, temperature, and composition. The definition could be applied to a portion of material in which the pressure, temperature, and composition varied from point to point, i.e., to a portion in disequilibrium: in this case, the value of the ratio μ_1 would depend on the details of the variation and on precisely where in the sample the addition δm_1 was made. More usefully, the definition is applied to a portion of material in equilibrium, i.e., with no internal variations of pressure, temperature or composition, and to addition processes that leave the sample still in equilibrium after the process is complete, i.e., reversible processes; in such cases, the value of the ratio μ_1 depends only on the current state of the sample and not on its history.

Thus,

$$\mu_1 = \left.\frac{\partial G}{\partial m_1}\right|_{P,T,m_2,\ldots,m_n} \tag{2.2}$$

and

$$dG|_{PT} = \mu_1\, dm_1 + \mu_2\, dm_2 + \cdots + \mu_n\, dm_n \tag{2.3}$$

The customary unit for G is the joule; m_1 can be expressed either in kilograms or kilogram-moles so that the units of μ are J/kg or J/kg-mole (or less commonly J/g, J/g-mole, J/m^3, kilocalories/g-mole, etc.).

To establish numerical values for μ, one has to pick arbitrarily a reference state and declare that for this state, $\mu = 0$. Consider the compound H_2O and the conditions 273 K and 1 bar: if water is taken as the reference state so that $\mu_{\text{water}} = 0$ kJ/kg-mole, then $\mu_{\text{steam}} = 8500$ kJ/kg-mole in these conditions. On the other hand, if the molecular gases H_2 and O_2 are taken as reference states, then on this basis $\mu_{\text{steam}} = -230\,000$ kJ/kg-mole. In the first instance, μ is positive; steam is less stable than the reference state and might spontaneously convert. In the second instance, μ is negative; steam is more stable than the reference state.

The difference in magnitudes is also significant. In the conversion *gases* $\rightarrow$ *steam* a reaction is involved, whereas the conversion *steam* $\rightarrow$ *water* is merely an aggregation process affecting molecules already present. If two alternative forms exist that are both solids, the difference in potential is typically smaller still, such as 2900 kJ/kg-mole for the pair diamond and graphite.

Properties of μ

The main properties we need to note about μ are the way it changes with pressure, temperature, composition, and the atomic structure of the material. (μ is also affected by the position of the sample in the local fields—gravitational, electrical, etc.—but we shall give no attention to that aspect.)

For pressure,

$$\frac{\partial \mu_1}{\partial P} = V_1 \qquad (2.4)$$

For example, if μ_1 is in J/kg and P is in Pa, V_1 is the partial volume of 1 kg of component 1 as it occurs in the material, in J/Pa-kg, i.e., in m^3/kg. Or if μ_1 is in J/mole, V_1 is the partial molar volume of the component in m^3/mole.

If we take the iron oxide *magnetite*, Fe_3O_4, as an example, values are:

relative density	5.2
volume of 1 kg	192 cm^3, about one cupful
molecular weight	232
for 1 g-mole,	mass = 232 g
	molar volume = 44 cm^3/g-mol
for 1 kg-mole,	mass = 232 kg
	molar volume = 0.044 m^3/kg-mol
for 1 kg-mole	$d\mu/dP$ = 0.04 J/kg-mol-Pa
	= 4 kJ/kg-mol-bar

For temperature,

$$\frac{\partial \mu_1}{\partial T} = -S_1 \qquad (2.5)$$

For example, if μ_1 is in J/kg and temperature is in K, S_1 is the partial entropy of 1 kg of component 1 as it occurs in the material, in J/kg-K. Or if μ_1 is in J/kg-mol, S_1 is the partial molar entropy in J/kg-mol-K. Again, magnetite serves as an example: at 298 K the entropy is 146 kJ/kg-mol-K. A change of temperature of 10 K gives a change in the chemical potential of 1460 kJ/kg-mol; to produce an equivalent change by increasing the pressure, the pressure increment needed is 365 bar or 36.5 MPa. In terms of everyday human experience, one might say that a modest and familiar change of temperature has effects that are matched only by a large and unfamiliar change of pressure, but that is because humans are familiar with only a very limited range of pressures. A change of pressure of 36.5 MPa is not uncommon in the working life of machine parts or structural components.

For composition, μ_1 goes down as the concentration of component 1 goes down. An equation is

$$\mu_1 = \mu_1^0 + RT \ln A_1 X_1 \qquad (2.6)$$

where μ_1^0 is the chemical potential of the same material when composed wholly of component 1, R is the gas constant, X_1 is the concentration of component 1, and A_1 is an "activity coefficient." Here X_1 is a fraction between 0 and 1, usually (moles of component 1)/(total moles in sample). $A_1 X_1$ also has maximum value 1 when $X_1 = 1$, i.e., when the material is composed wholly of component 1, so that the variation of μ is as shown in

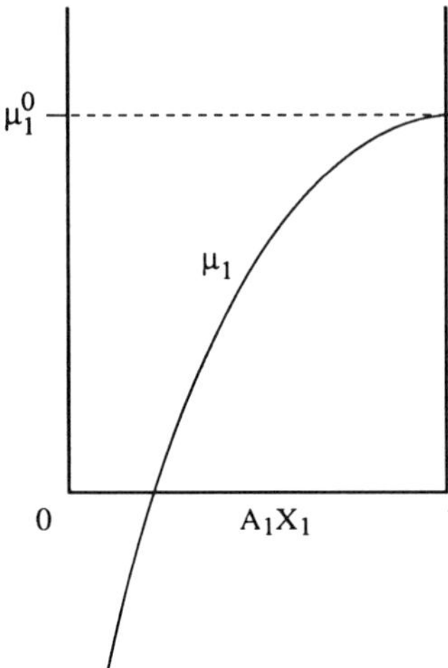

Figure 2.1 Variation of chemical potential of component 1 with the component's activity A_1X_1.

Figure 2.1. But when X_1 is less than 1, the value of A_1 is rather unpredictable and has to be discovered, for example, by experiments. Hence eqn. (2.6) does two things: it embodies the expectation that as X_1 goes down, the chemical potential μ_1 goes down, and *strongly so* as X_1 approaches zero (an expectation of a qualitative rather than quantitative kind); and it defines the multiplying factor A_1 that is customarily used to make experimental estimates of μ_1 lie on a logarithm-shaped curve. For the purposes of these chapters, the qualitative trend is the part to emphasize.

Variation of μ with atomic structure. Some materials are polymorphic; examples are α- and γ-iron, α- and β-quartz, and the graphite and diamond forms of carbon. The chemical potential has a different value in each polymorph; for example, at 298 K and 1 bar pressure, the values for graphite and diamond are:

Graphite: 0   diamond: 2900 kJ/kg-mol
(molar volume 5300 cm^3/kg-mol)   (molar volume 3400 cm^3/kg-mol)

We can use these numbers to estimate chemical potentials at higher pressures. If we take the molar volumes as constant, a change ΔP in pressure gives an increase $V \cdot \Delta P$ in μ (from eqn. 2.4). For example at 1001 bars, $\Delta P = 1000$ bar $= 100$ MPa, so that

$$\Delta\mu = 530 \text{ kJ/kg-mol for graphite (new value of } \mu = 530 \text{ kJ/kg-mol)}$$

and

$$\Delta\mu = 340 \text{ kJ/kg-mol for diamond (new value of } \mu = 3240 \text{ kJ/kg-mol).}$$

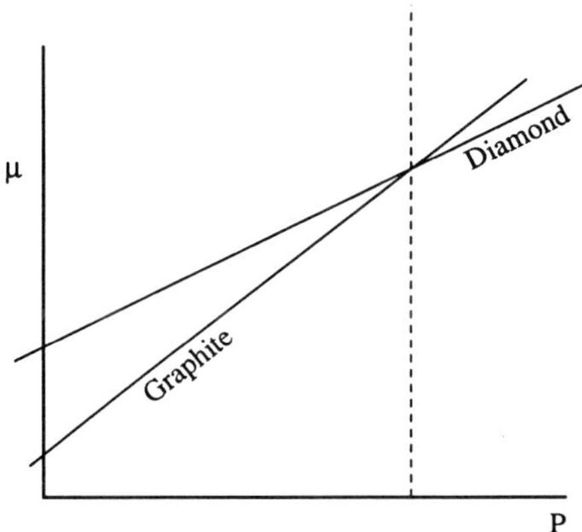

Figure 2.2 Variation of chemical potential of carbon with pressure.

A diagram also serves; see Figure 2.2. It can be estimated that the two forms would have equal μ at 1530 MPa; above this pressure, μ(diamond) < μ(graphite) and diamond would be the stable form.

The atomic structure can be different, as just noted, in two polymorphs; it can also be different in two polymers. Vinyl chloride, CH_2CHCl, for example, can exist as a monomer—each molecule separate from its neighbors; or in dimer form, i.e., molecules paired; or in 100-mer form, with the monomers polymerized into chains of 100 each. Most polymers are, of course, mixtures of chain lengths—it is difficult technologically to get them all exactly the same; but the notion serves to make the point that the chemical potential per mole or per kilogram can be different according to the state of polymerization the kilogram of material is in. The monomer and the dimer are clearly distinct states, like graphite and diamond, but at higher degrees of polymerization the different possibilities are less distinct. A simple view to take is that a polymer sample has a *mean molecular weight* that can vary in a *continuous* way, and that the chemical potential per kilogram is a smooth function of the mean molecular weight; see Figure 2.3. But this view is slightly fuzzy, not having the definiteness of the graphite–diamond example, as follows.

Nonequilibrium states

Along with the definition of chemical potential, it was stated that the definition is most useful when applied to materials in equilibrium. Once one envisages samples is disequilibrium, there are so many possible states and possible histories leading to them that one might ask: is trying to define a chemical potential any use? We avoid giving a definite answer to this question right away; it recurs in the following chapters and eventually gets

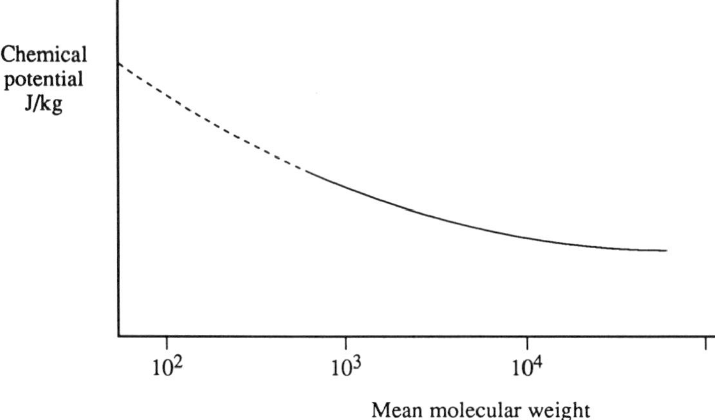

Figure 2.3 Variation of chemical potential with degree of polymerization. The quantity plotted is expressed in joules per kilogram rather than in joules per mole.

a tentative answer "Yes, sometimes." For the present, we simply note two facts that bear on the eventual answer.

The first fact to note is that Figure 2.2, for graphite and diamond, implies that potentials can be assigned even in disequilibrium. Strictly, the μ,P diagram should be as in Figure 2.4. But obviously, diamonds exist outside their stability field, and extrapolating the line of μ-values permits useful predictions about how they will behave; the state "diamond" continues to be so well defined that we are comfortable assigning a potential to it. (In fact the thing that is mentally difficult is to keep alive the idea that any diamond we see is in process of spontaneously converting to graphite.)

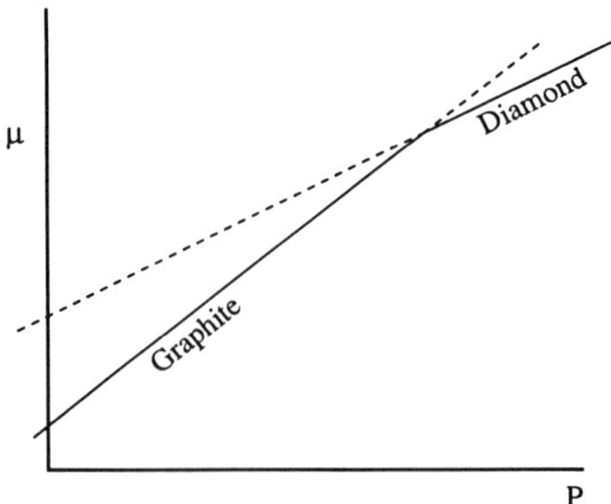

Figure 2.4 Variation of chemical potential of carbon with pressure. Figure 2.2 has been revised to distinguish metastable states from stable states.

The second fact is that what goes for diamond and graphite goes also for the monomer and dimer of vinyl chloride: at prescribed P and T, either one form is the more stable or the other is. But then we can shift mentally to a condition where the molecules polymerize spontaneously: the 100-mer form is more stable than either the monomer or the dimer, and the 1000-mer form is slightly more stable yet. But probably none of these is the ultimate or ideal stable form, whatever that may be; the values that Figure 2.3 shows are all disequilibrium values. Yet the diagram seems to contain some real, possibly useful, information.

The only conclusion needed at this point is a cautious one: for a component in a sample in true equilibrium, a chemical potential can certainly be defined, and it seems that in certain circumstances, even where we know that the material is not in its ultimate equilibrium state, the idea of a component having a chemical potential continues to be acceptable and perhaps for some purposes useful.

3
DISEQUILIBRIUM 1: POTENTIAL GRADIENTS AND FLOWS

As in Chapter 2, so again here the intention is to review ideas that are already familiar, rather than to introduce the unfamiliar; to build a springboard, but not yet to leap off into space.

Gradient of pressure or concentration

The familiar idea is of flow down a gradient—water running downhill. Parallels are electric current in a wire, salt diffusing inland from the sea, heat flowing from the fevered brow into the cool windowpane, and helium diffusing through the membrane of a helium balloon. For any of these, we can *imagine* a linear relation:

Flow rate across a unit area = (conductivity) × (driving gradient)

where the conductivity retains a constant value, and if the other two quantities change, they do so in a strictly proportional way. Real life is not always so simple, but this relation serves to introduce the right quantities, some suitable units and some orders of magnitude.

For present purposes, the second and fourth of the examples listed are the most relevant. To make comparison easier we imagine a barrier through which salt can diffuse and through which water can percolate, but we imagine circumstances such that only one process occurs at a time. Specifically, imagine a lagoon separated from the ocean by a manmade dike of gravel and sand 4 m thick, as in Figure 3.1. If the lagoon is full of seawater but the water levels on the two sides of the dike are unequal, *water* will percolate through the dike, whereas if the levels are the same and the dike is saturated but the lagoon is fresh water, *salt* will diffuse through but there will be no bulk flow of water. (More correctly, because seawater and fresh water have different densities, and because of other complications, the condition of no

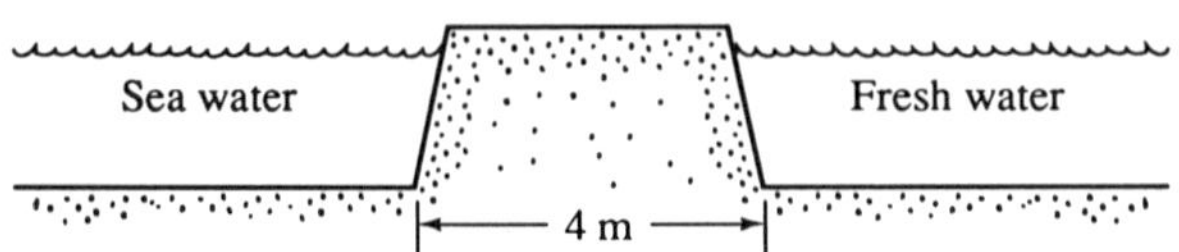

Figure 3.1 A permeable barrier between two bodies of water.

Table 3.1 Suitable units for transport quantities.

	flow rate	gradient	conductivity factor
water flow	kg/m^2-sec m^3/m^2-sec	Pa/m	kg/Pa-m-sec m^3/Pa-m-sec or m^2/Pa-sec
salt flow		$\dfrac{d(\text{concentration})}{d(\text{distance})}$	
	atoms/m^2-sec	$\dfrac{\text{atoms/m}^3}{\text{m}}$	m^2/sec
	moles/m^2-sec	$\dfrac{\text{moles/m}^3}{\text{m}}$	
	kg/m^2-sec	$\dfrac{\text{kg/m}^3}{\text{m}}$	

net water flow would be achieved in circumstances a little different from what was just stated. For present purposes all we need is the idea that conditions exist where water does not percolate but salt does diffuse.) For flow of water driven by a pressure gradient, suitable units are shown in the upper part of Table 3.1 and for diffusion of salt driven by a concentration gradient, suitable units are shown in the lower part.

Numerical examples follow. First, for flow of water, let the material of the dike have a conductivity factor or transmissivity of 2×10^{-5} kg/Pa-m-sec and let the difference in pressure from one side to the other be 0.008 MPa (for example, because of a difference in water level of about 30 inches or 75 cm). The average pressure gradient would be 0.002 MPa/m and the flow rate would be 0.04 kg/m^2-sec—a cupful in 5 seconds for every square meter of cross-section area; material with such a high transmissivity would not be suitable for dike purposes.

Second, for diffusion of salt, let the material of the dike have a conductivity factor or diffusivity of 5×10^{-9} m^2/sec. This is a value for the wet sand en masse regardless of the geometry of the tiny water pathways that permeate it. The salt content of seawater is about 3 g per 100 g or 3 kg per 100 kg; a cubic meter of seawater has a mass of about 1100 kg so the salt content is about 33 kg/m^3; if the concentration changes linearly from the salty side to the freshwater side, the change is 8.3 kg/m^4; then the diffusion rate for salt is 4×10^{-8} kg/m^2-sec. A year is about 3×10^7 sec, so that this equals 1.2 kg or nearly 3 lb of salt crossing per square meter per year.

To express the same result using moles, take the molecular weight of NaCl as 60 so that 1 kg-mole = 60 kg. Then the concentration in seawater is 33/60 or 0.55 kg-mol/m^3 and the diffusion rate is 0.02 kg-mol/m^2-year.

The purpose of the examples is to bring three points to attention as follows.

1. For flow of water, the rate of $0.04\,\text{kg/m}^2$-sec could be expressed as $36\,\text{cm}^3/\text{m}^2$-sec or $36 \times 10^{-6}\,\text{m}^3/\text{m}^2$-sec or $36 \times 10^{-6}\,\text{m/sec}$ or 0.036 mm/sec, but nothing actually moves at this velocity. If the water channels among the sand grains constituted 3 percent or 0.03 of an average cross-sectional area, the actual average velocity of the water would be 1.2 mm/sec, but for present purposes the overall mass or volume transmitted per m^2-sec is the quantity of interest, regardless of the internal details.

2. The conductivity factor for salt diffusion has deceptively simple units. If we express the concentration gradient in mol/m^4 and the transport rate of salt in kg/m^2-sec, the conductivity factor has units

$$\frac{\text{kg}}{\text{m}^2\text{-sec}} \bigg/ \frac{\text{mol}}{\text{m}^4}$$

It is only when concentration and transported mass are expressed in the same units, such as moles or kilograms, that the factor appears as m^2/sec. Though less elegant, the more complicated expression of the factor's units is a more complete indication of what the factor represents.

3. The item here called a conductivity factor has various names—permeability, diffusivity, etc.—that sometimes emphasize the host material (e.g., "permeability of sandstone") and sometimes emphasize the traveling material (e.g., "diffusivity of hydrogen"). The factor in reality always depends on both host and traveler; it is a property of the transport situation as a whole. Sometimes it is useful to separate out two components such as "mobility of the diffuser" and "tortuosity of the matrix" but for present purposes we shall stay with a single comprehensive factor. The terms *permeability* and *diffusivity* may be used from time to time, but we shall try to maintain the view that any conductivity factor is acceptable, under whatever name, as long as its units are clearly in view.

Gradients in potential

The first example, flow of water, can readily be rewritten in terms of potentials. Equation (2.4),

$$\frac{\partial \mu_\text{B}}{\partial P} = V_\text{B} \qquad (\text{where B} = \text{name of component considered})$$

shows that, if pressure P is the only variable, an overall change from state 1 to state 2 is described by

$$\mu_1 - \mu_2 = \int V_\text{B}\, dp \qquad (3.1)$$

And, to whatever extent it is permissible to treat V_B as independent of pressure, this can be rewritten

$$\mu_1 - \mu_2 = V_B(P_1 - P_2) \tag{3.2}$$

Here μ_1 is the equilibrium potential of component B in the first equilibrium state and μ_2 is the equilibrium potential in the second equilibrium state. Equation (3.2) strictly says nothing at all about the situation in Figure 3.1, because it is not an equilibrium situation. But one can suggest that the water on one side is in every respect *very similar* to water in equilibrium at one pressure and the water on the other side is *very similar* to water in equilibrium at a second pressure. Hence, there is a potential drop over the 4 m interval that is *very close* to (0.008 MPa) × (volume of 1 mass unit of water). One kilogram of water occupies 10^{-3} m^3 and one kg-mol occupies 18×10^{-3} m^3. Taking a kilogram as the mass unit, the potential drop is 8 Pa-m^3/kg or 8 J/kg; using kg-mol, the potential drop is 144 J/kg-mol. These give potential gradients of 2 J/kg-m or 36 J/kg-mol-m. To work, for example, with the first of these we would have to construct a new conductivity factor for the material of the barrier, which would now be written as 0.02 kg^2/J-m-sec; we would continue to predict (by multiplication) a flow of 0.04 kg/m^2-sec. As before, the units of the conductivity factor are more intelligible if written as

$$\frac{\text{kg}}{\text{m}^2\text{-sec}} \bigg/ \frac{\text{J/kg}}{\text{m}}$$

flow rate   potential gradient

or, in molar amounts

$$\frac{\text{kg-mol}}{\text{m}^2\text{-sec}} \bigg/ \frac{\text{J/kg-mole}}{\text{m}}$$

We will now try to make a similar conversion in the example with a concentration gradient. The appropriate equation is eqn. (2.6):

$$\mu_B = \mu_B^0 + RT \ln A_B X_B$$

or

$$\mu_B = \mu_B^0 + RT \ln \alpha_B$$

if we use α_B, the *activity* of component B, to stand for the product $A_B X_B$. If two states of a material containing B differ only in their concentrations of B, then μ_B^0 is the same for the two states. Then if, as before, we designate the states by subscripts 1 and 2,

$$\mu_1 - \mu_2 = RT \ln \alpha_1 - RT \ln \alpha_2 = RT \ln \frac{\alpha_1}{\alpha_2}$$

If α_1 and α_2 differ only slightly, both being close to a common value α, then

$$\mu_1 - \mu_2 = \frac{RT(\alpha_1 - \alpha_2)}{\alpha} \tag{3.3}$$

Comparing eqns. (3.3) and (3.2), and recalling that the factor V_B converted a pressure gradient to a potential gradient, we suppose that the factor RT/α_B will similarly convert an activity gradient to a potential gradient. In the example discussed, let the activity coefficient for salt be 0.91 and let the gradient in activity be linear across the 4 m of dike. Then the concentration 0.55 kg-mol/m^3 on the salty side gives an activity gradient of 0.13 kg-mole/m^4. The factors R and T can be taken as 8320 J/kg-mole-K and 300 K, respectively, and $\alpha = 0.50$ kg-mol/m^3, so that the potential gradient is $(0.63) \times 10^6$ J/kg-mole-m. If the resulting flux were indeed 0.02 kg-mol/m^2-year or $(0.7) \times 10^{-9}$ kg-mol/m^2-sec, we would conclude that the conductivity factor relating flux of salt to potential gradient was $(1.1) \times 10^{-15}$ (kg-mol)^2/J-m-sec.

The numbers just given are not unreasonable for a small region in a neighborhood where $\alpha = 0.50$ kg-mol/m^3; but clearly the calculation begins to look strange if we try to pursue it in a region near the freshwater side of the barrier, where α approaches zero. At this side, the potential gradient becomes enormous but the flux stays steady. Does the conductivity factor for some reason become very small? It is more reasonable to suppose that the conductivity factors stays roughly constant, but *the number of atoms the driving gradient acts on becomes very small.*

At this point, we note a peculiarity of the suggestions made so far: it was suggested that a linear gradient in activity of 0.13 kg-mol/m^4 gave a flux of 0.02 kg-mol/m^2-year across any representative square meter of barrier, both toward the salty side where salt is abundant and toward the freshwater side where far fewer salt ions are present. This implies that the average velocity of the ions is small on the salty side and vastly larger on the freshwater side. In short, the net flux is proportional to the activity gradient but the particle velocity is proportional to the potential gradient; we have two linear relations operating simultaneously:

$$\text{Flux} \quad = \quad K_1 \quad \times \text{ activity gradient} \tag{3.4a}$$
$$\text{mole/m}^2\text{-sec} \quad \text{m}^2\text{/sec} \quad\quad \text{mole/m}^4$$

$$\text{Particle velocity} = \quad K_2 \quad \times \text{ potential gradient} \tag{3.4b}$$
$$\text{m/sec} \quad\quad \text{mole-m}^2\text{/J-sec} \quad\quad \text{J/mole-m}$$

Also,

$$\text{Potential gradient} = \frac{RT}{\text{activity}} \times \text{activity gradient} \qquad (3.4c)$$

$$\text{Flux} = \text{particle velocity} \times \text{concentration} \qquad (3.4d)$$

$$= (K_2 \times \text{potential gradient}) \times \text{concentration} \quad (3.4e)$$

$$K_1 = K_2 \times RT \times \frac{\text{concentration}}{\text{activity}} \qquad (3.4f)$$

This is a fully coherent set of equations, so that eqns. (3.4a) and (3.4b) are "equally true." But one might suggest that (3.4b) is the more fundamental, and the simplicity of (a) is a little fortuitous, being a neat consequence of eqns. (3.4b), (3.4c) and (3.4d).

Working with the details of the equations should not obscure three salient points:

1. As with pressure gradients, so with concentration gradients, the linear relation first put forward can be replaced by a slightly different linear relation in which the material response, the flux, is linked to a gradient in chemical potential.
2. Making this shift of emphasis is a step toward unification, toward our being able to handle problems where there are gradients in pressure and concentration acting simultaneously.

But

3. To use the concept of chemical potential in contexts where material is being driven from point to point, i.e., in nonequilibrium situations, involves an approximation or assumption. As a procedure, it has proved successful in 30 or 40 years of use, but we should look a little more closely in the next chapter at the nature of the approximation made.

APPENDIX 3A: DIFFUSION AT AN INTERFACE

The purpose of this appendix is to present an idea that is used in Chapter 16. The topic is not on the main line of progress from Chapter 3 to Chapter 4 and one option is to pass over it. On the other hand, the topic requires only ideas from Chapters 2 and 3 for its discussion; in this sense, it belongs early in the sequence of topics treated, and the other option is to regard it as extra practice with those ideas. As elsewhere, some prior knowledge is assumed.

We consider a compound such as (Fe, Zn)S. It exists in two forms, distinguished by cubic symmetry or hexagonal symmetry, and in either form the ratio of Fe to Zn is variable. There is a range of pressures and

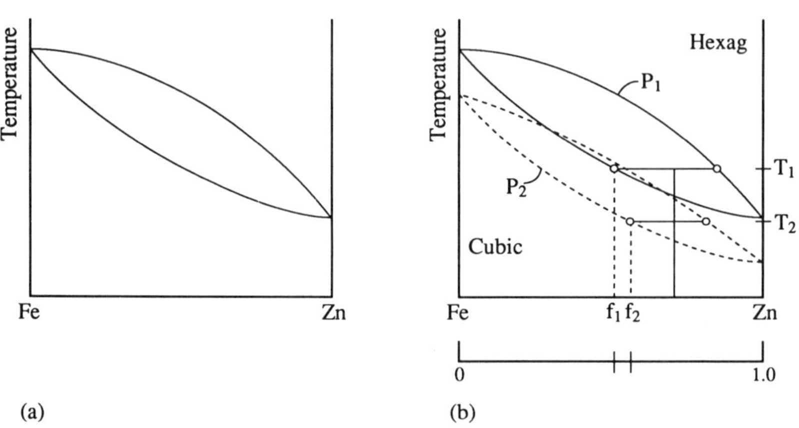

Figure 3.2 Phase boundaries in materials of composition (Fe, Zn)S.

temperatures where the two forms can coexist in equilibrium. When they coexist, the Fe/Zn ratio takes different values in the two forms, as in Figure 3.2a.

The coexisting forms are affected by change of pressure as well as change of temperature. Equilibria at two different pressures are shown in Figure 3.2b and attention is given to the following: it is possible for 1 kg to exist as 400 g of cubic form and 600 g of hexagonal form either at P_1,T_1 or at P_2,T_2. Indeed there is a path from P_1,T_1 to P_2,T_2 through a series of states in any of which the equilibrium state of the kilogram is a partition into 600 g of one form and 400 g of the other. Moving to and fro along this path does not provoke any change from cubic form to hexagonal, but it does provoke exchange of Fe and Zn between the two forms, as discussed next.

In the particular conditions of Figure 3.2b, at P_1,T_1 the mole-fraction of Zn in the cubic phase is f_1 and at P_2,T_2 the mole-fraction is f_2. If equilibrium is established at P_1,T_1 and conditions are then jumped suddenly to P_2,T_2 and thereafter kept steady, at any site where the two phases are in contact an evolution will occur as in Figure 3.3a: right at the interface, adjustment of the phase compositions by exchange of Fe and Zn is instantaneous, establishing the new value f_2, but in the interior of either phase, adjustment of composition can occur only at such rate as interdiffusion of Fe and Zn permits. After a short time, a steep profile of composition with distance will exist and after a longer time the profile will become flatter; in an ideal experiment, it would take an infinite time for the cubic phase to reach composition f_2 uniformly throughout its whole extent.

But we wish to consider a gradual progress from the first state *toward* the second, rather than a sudden jump. To do this, let the path from start to finish be divided into one hundred steps. We jump the first step and then wait while the composition profile acquires a certain breadth; then we jump

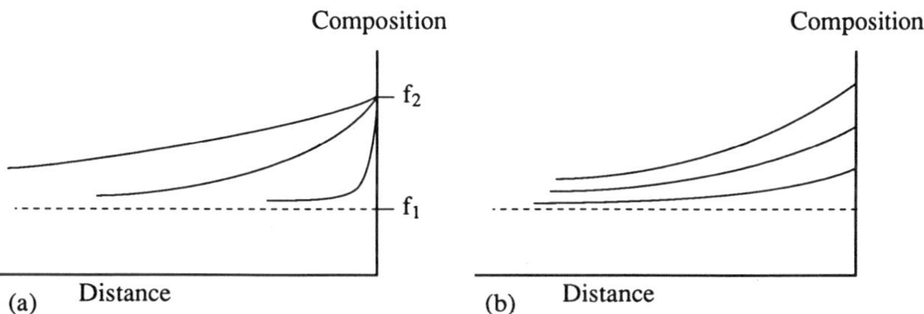

Figure 3.3 Composition profiles at an interface. In (a), all three curves have the same peak height (at the interface) but different breadths. In (b), all curves have different heights but the same breadth, in the following sense: if the peak height of any curve is h, the distance horizontally from the interface to the point where the curve has height $h/2$ is the same for all curves in the set, and similarly for any other fraction of h.

the next step and wait again. The purpose of imagining such a series of steps and wait-periods is to establish the possibility of keeping the breadth of the diffusion profile constant while gradually increasing its gradient, as in Figure 3.3b. In the first history, the peak height is kept constant and the profile's breadth changes; in the second history, the profile's breadth is kept constant while the peak height changes. The first history could, as noted, be followed for an infinite time, whereas the second cannot continue indefinitely; processes could run as described for a certain amount of time but eventually would have to switch to some other style of change.

The discussion just given is intended partly as review of some needed ideas about equilibrium, coexisting phases, and diffusion processes; but the particular history shown in Figure 3.3b is discussed again in Chapter 16 and this appendix can be referred to in that connection.

4

DISEQUILIBRIUM 2: ASSOCIATED EQUILIBRIUM STATES

The purpose of this chapter is to consider the question: how much uncertainty, or how coarse an approximation, is involved when we use ideas borrowed from equilibrium thermodynamics and apply them in nonequilibrium situations?

To be specific, consider eqn. (3.2):

$$\mu_1 - \mu_2 = V_B(P_1 - P_2)$$

When we have in mind two equilibrium states, isolated from each other, one static at pressure P_1 and the other static at pressure P_2, the equation is a regular part of equilibrium thermodynamics and has a high degree of validity on that account. But when we have in mind a single nonequilibrium state, part of which is almost static at pressure close to P_1, and part of which is almost static at pressure close to P_2 (and part of which is not so close to static at all), is there really a contrast in chemical potential between one portion of the material and another portion of the same material farther along the system? And if so, can it really be estimated by using eqn. (3.2)?

The answers, fortunately, are *Yes* and *Yes*. For an example where the nature of the approximation we make can be quite clearly seen, consider the long bar in Figure 4.1.

Diffusion in a long bar

The bar is uniform in composition except for one minor constituent R whose concentration diminishes linearly along the bar from left to right (and for purposes of this example we assume that the activity coefficient A is exactly 1 everywhere so that R's activity and its concentration are the same); the concentration is represented by the sloping line in Figure 4.1b. We arbitrarily select five small portions of the bar for consideration; for any single portion, an average concentration of R can be established, and we can *imagine* a material that is homogeneous, with that concentration of R at all points. In such a material, the chemical potential of R is perfectly well defined. In what ways does the portion of the actual bar differ from a portion of this imagined homogeneous material?

In the first place, the left-hand end of the real portion is slightly above average in R and so resembles a material with slightly higher chemical

30

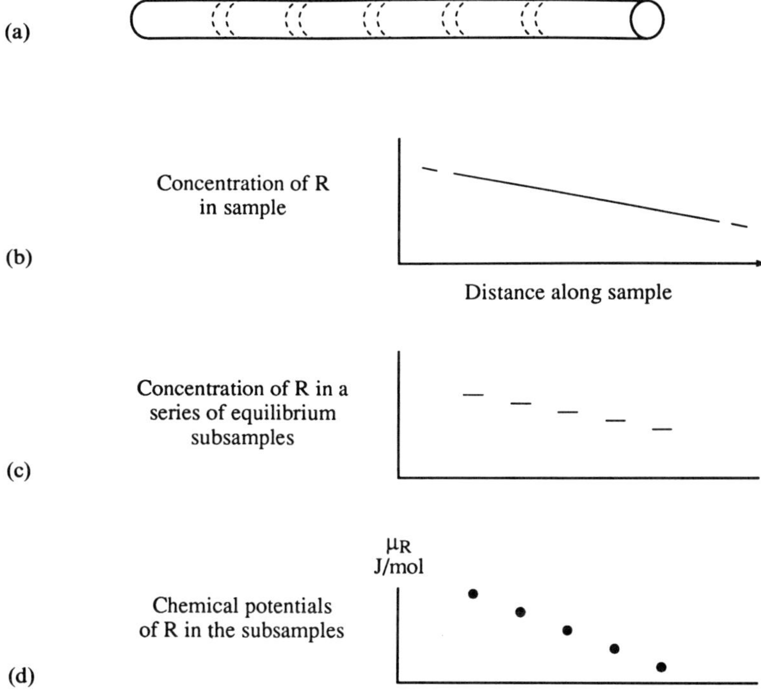

Figure 4.1 A bar with nonuniform composition and the properties of some associated imaginary states.

potential; but the right-hand end is deficient in R to just the same extent, so to a first order of approximation the average or integrated effect of the $RT \ln X$ term is the same in the real portion and the imagined homogeneous portion. But to a second order of approximation it is *not quite* the same.

Figure 4.2a shows an ideally linear profile of concentration X. Then if X_0 is the mean value of X for the portion of bar shown, the two triangles have exactly equal areas. But the entropy of the sample is controlled by $R \ln X$ as shown in Figure 4.2b. If the profile of X is linear, the profile of $\ln X$ is curved; the areas of the two stippled portions are now not equal and the integrated area under the curve is less than that under the horizontal line. In more direct terms, if we take a portion of bar in which component R is homogeneous and push some of the R up to one end, we are imposing a degree of order on R—we are decreasing its entropy; the entropy of R in a graded concentration is slightly less than in a uniform concentration of the same total quantity in the same space.

But having seen in Figure 4.2b a quantitative picture of how a graded concentration has lower entropy than a uniform concentration, we see also that in a sample that is sufficiently small, the difference is negligible. In

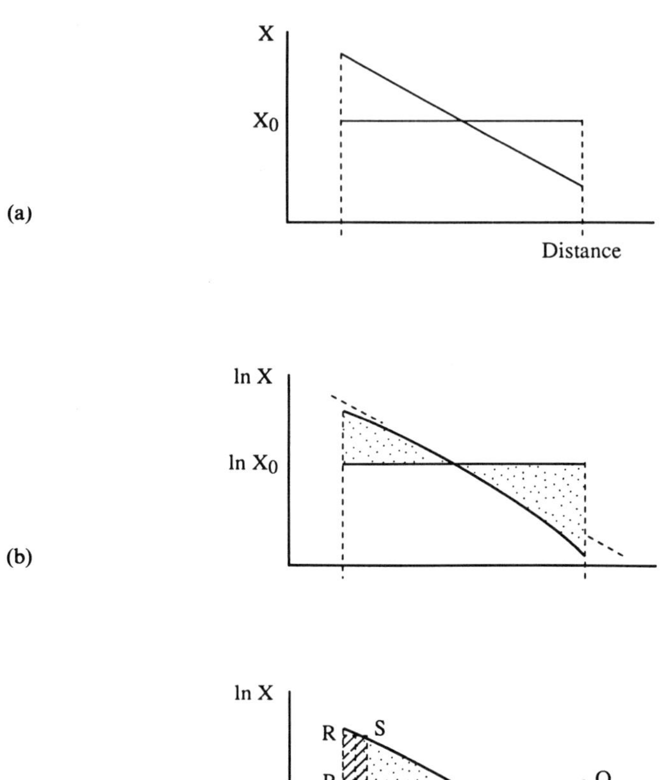

Figure 4.2 Variation of X and $\ln X$ with position in a nonuniform bar; X is the concentration of the component of interest.

Figure 4.2c, the stippled area is not well represented by the area under line PQ but the shaded area is well represented by the area under the line RS.

Summary. For any small portion of the graded bar, we can imagine a portion of the same size and overall R-content, in which R is homogeneous instead of being graded. The imagined sample is in equilibrium; its state is the *associated equilibrium state* that is associated with the disequilibrium portion we decided to consider.

A distinction can be made at this point that is important in later chapters. It is not suggested in the reasoning just given that the small graded sample is in any sense in equilibrium. The essential feature of a gradient is that it retains a well-defined nonzero value as the length of line considered goes to zero; a tiny sample of graded composition is out of equilibrium to just as great an extent as a larger sample of the same gradation. The useful property of the small sample is not "local equilibrium"; its useful property is that its total entropy is the same as an easily imagined sample that is in equilibrium. Local equilibrium is not needed for the reasoning used nor does absence of local equilibrium make the reasoning invalid. Continuity is all that is needed in the properties of the real gradational sample.

Associated equilibrium states

Does every small portion of a nonequilibrium system have an associated equilibrium state? And if so, how do we discover it? According to de Groot's prescription (1951, p. 11) we imagine the nonequilibrium portion to be suddenly isolated from its surroundings in all respects and allowed to grind to a halt—all gradients flattened and all rates of processes gone to zero: the state so reached is the associated equilibrium state for the portion of nonequilibrium system considered.

This agrees with the description of the graded bar and the equilibrium states we imagined for various portions of that; and the approach can clearly be used for the salt-diffusion problem, at the 4 m-wide dike. To treat pressure differences in a parallel manner, one notes that materials become less compressible when compressed; the pressure–volume curve for a fixed mass is usually concave away from the origin, as a hyperbola is. Then a profile of graded pressure is like a profile of graded concentration: it is intrinsically and unquestionably a nonequilibrium state at every point, but the properties of any small portion can be matched with the properties of an imagined equilibrium state.

The idea of an associated equilibrium state is thus quite general: it is a very powerful concept that can be used more accurately and more confidently than the concept of local equilibrium. Confidence comes from two sources: first we can see that the real nonequilibrium state and the imagined equilibrium state are closely similar, and second, we can bring the exact differences to light—even, in some circumstances, express them quantitatively and demonstrate their smallness. It is the second aspect that brings confidence, and allows us to use associated equilibrium states to define potentials and predict material responses, even in situations where the exact magnitude of the small differences cannot actually be worked out in numbers.

Indeterminacy and dependence on history

This section is off the main trend. The suggestion is sometimes made that it is *only* for equilibrium states that a chemical potential can be defined, and that for nonequilibrium states there is no chemical potential—the concept does not exist. A related suggestion is that, for a nonequilibrium state, a component's chemical potential exists but has a value that is affected by the sample's history in ways we can never follow out, so that the chemical

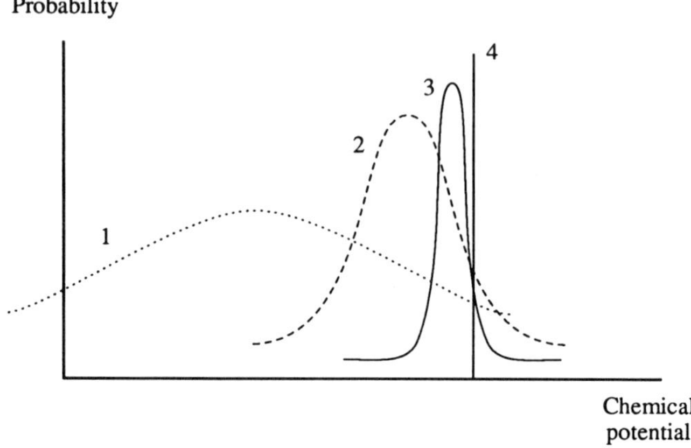

Figure 4.3 Different degrees of definiteness in a sample's chemical potential.

potential is indeterminate and not available as a guide in predicting behavior. Readers willing to brush these suggestions aside can skip on to Chapter 5; without question, *something* drives a material's behavior in nonequilibrium situations, and predictions succeed if based on the idea that that something is approximated by the material's associated equilibrium potential. So one can proceed without worrying whether the chemical potential in a nonequilibrium state is fictitious or not. However, for those of us who are inclined to worry, Figure 4.3 is offered and discussed. (Beside being off the main trend, the following suggestions are more speculative and less well-founded than the main body of the text.)

Figure 4.3 shows four curves in a series and the reader is asked to imagine curves 5, 6 and 7 resembling curves 3, 2 and 1 but fading away to the right. The sequence 1 through 7 is a sequence from a strongly nonequilibrium state 1 through diminishing degrees of disequilibrium to an equilibrium state, 4, and then on into nonequilibrium again. If the pressures on two sides of a barrier like the dike in Chapter 3 were $P + x$ and $P - x$, and we observed the barrier's centerpoint while x went through a series of decreasing values $x_1, x_2, x_3, 0, -x_3, -x_2, -x_1$, the centerpoint would experience a sequence of states of the intended type. Ideas to emphasize are, first, that the progression could be made totally continuous: the unique equilibrium state, $x = 0$, is part of a totally smooth progression of values of x; isolating seven discrete values is purely for the purposes of the diagram. Second, in the equilibrium state, the material's chemical potential is unquestionably and perfectly well defined. Figure 4.3, then, is a sketch of how a quantity can progress smoothly from being very ill-defined through states of being fairly well defined; momentarily, it is absolutely defined and then gradually becomes ill-defined again.

A feature of the diagram is that even curve 3 has tails going off to infinity;

that is to say, as soon as the system departs from equilibrium, it enters a state where the chemical potential can have absolutely any value. On the other hand, the probability is that its value lies in a range that is both narrow and close to curve 4. We thus take up a sort of compromise position: in a nonequilibrium situation, a component's chemical potential is indeterminate and can have any value; but if the departure from equilibrium is not too gross, the value *probably* lies in a narrow range and the center of its range is only slightly off the equilibrium value. Thus the associated equilibrium state is the best predictor that comes to hand, and is likely to be a good predictor, despite the fact that a large degree of indeterminacy theoretically exists.

5

DISEQUILIBRIUM 3: INTERNAL VARIABLES

In Chapters 2, 3, and 4, the usefulness of the concept *chemical potential* has been explored for describing and predicting movement of material from point to point in space—from a location where a component's potential is high to a location where its potential is lower. But chemical potential influences another type of material behavior as well, as in the example at the end of Chapter 2, the polymerization of vinyl chloride. The polymerization is a process that runs at a certain rate, like diffusion of salt, and the rate depends on the potential difference between the starting state and the end state; but unlike diffusion of salt, there is no overall movement from one location to a new location—the vinyl chloride simply polymerizes where it is. There are movements, of course, on the scale of the interatomic distances, but nothing corresponding to the 4 m of travel that appears in the discussion of the dike. If no *travel* is involved, it is not so easy to calculate a potential gradient along the travel path and go on to predict a rate of response. Yet there definitely is a rate of response, even with PVC polymerizing. The purpose of this chapter is to consider this matter; we shall then be equipped to begin considering nonhydrostatic conditions.

An example

The essential idea is to represent all possible degrees of polymerization along an axis, as in Figure 5.1. The figure is drawn to represent a condition where the chemical potential per kilogram is greater in the monomer form than in the dimer form, i.e., a condition where the material polymerizes spontaneously. Suppose we know the chemical potential per kilogram for all degrees of polymerization and also, at some temperature, the rates at which 2 forms from 1, 3 forms from 2, etc. (per kg of the starting form in a pure state). Then we arbitrarily pick a distance on the horizontal axis to separate point 1 from point 2. Thereafter, an arithmetic process can create a distance from point 2 to point 3 such that

$$\frac{\text{Gradient } 2 \rightarrow 3}{\text{Gradient } 1 \rightarrow 2} = \frac{\text{rate of conversion } 2 \rightarrow 3}{\text{rate of conversion } 1 \rightarrow 2}$$

and so on all along the axis.

The variable we are tracking, the degree of polymerization, is an *internal*

36

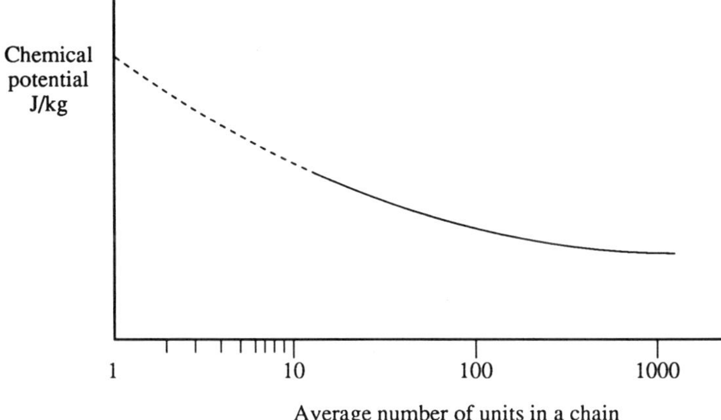

Figure 5.1 Variation of chemical potential with degree of polymerization.

variable (as opposed to a spatial variable in geographical space) and the artificial axis is an internal-variable axis.

The separation of two points on this artificial axis is a measure of the resistance to the conversion of material from the state represented by one point to the state represented by the next: if the distance is large it represents a difficult conversion, such that, even when the potential difference between the states is large, the conversion still does not run very rapidly. An example of a difficult conversion would be one where two molecules combine but only if one collides with the other at the right point and also at the proper orientation; even if this combination-process resulted in a large drop in potential, it would not run rapidly because of the unfavorable probabilities. By contrast, an easy conversion would be one that is not sensitive to the position where one molecule strikes the other or to orientation. Thus distances on the artificial axis represent genuine information about the material's behavior.

A significant difference should be noted between distances on an internal-variable axis and distances on a spatial axis, like the horizontal axis in Figure 4.1; the difference is that, as so far discussed, the internal-variable distances always fit the experimental data. They are created from the experimental data and are simply a graphical display of information that was inherent in the data before the internal-variable axis was ever created; we have not yet used the internal-variable idea to predict a result that could be *tested* by experiment. By contrast, a normal spatial axis can be used at once for prediction: if we made a barrier as in Figure 3.1 but 8 m wide, we would predict right away that the flow rate would be halved.

If we knew enough about the process and geometry of molecular collisions, of course, we could calculate the probabilities referred to above; then the first observed experimental rate of conversion would serve to test the success of the hypotheses in hand. But for many internal variables, considerable experimental data is "consumed" in setting up the variable, before it can ever be used to make a prediction.

Summary. Chemical-potential differences drive material from location to location through geographical space, but also drive conversion of material from one *state* to another without change of location. If a series of states differ one from the next by gradual change of a single attribute, this attribute is an *internal variable*; different states of the material can be represented by points on an *internal-variable axis*, and a diagram can show a gradient of chemical potential with respect to position on the internal-variable axis, just as another diagram can show a gradient with respect to position on a geographical space axis. A linear relation can be *sought* between the potential gradient with respect to an internal variable and the material's rate of change through successive values of the variable, but will not necessarily be found; different internal variables have different degrees of success—i.e., different degrees of usefulness in helping us to summarize behavior trends and make predictions.

Internal and spatial variables combined

To consolidate ideas introduced so far, consider the possibility of forming a PVC polymer of low molecular weight, so that it is a somewhat viscous fluid. We bring the fluid to a temperature where it tends to polymerize and pump it through a pipe; then different portions of fluid have different energy content through being at different *pressures* (at different positions along the pipe) and also different portions have different energy content through being at different *degrees of polymerization*. The situation can be summarized in a diagram showing a potential surface, as in Figure 5.2. According to the

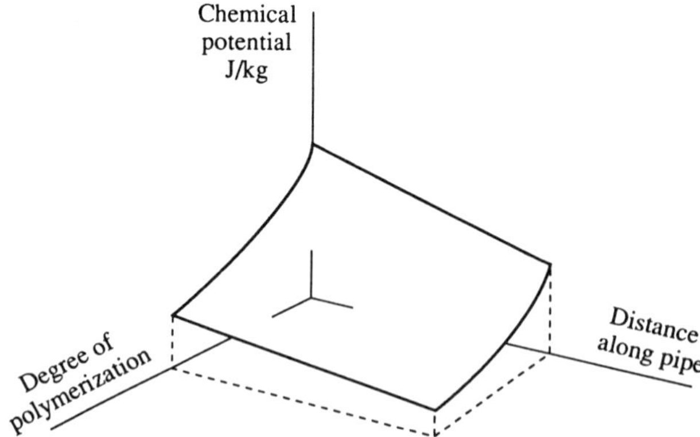

Figure 5.2 Variation of chemical potential with two parameters simultaneously.

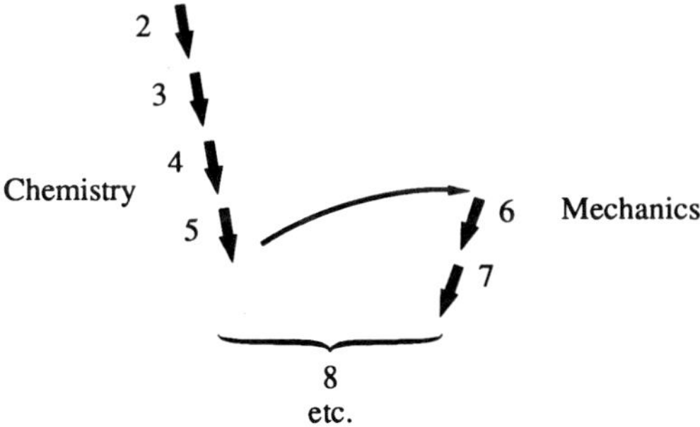

Figure 5.3 A flow chart showing the sequence of chapters.

gradients at different points on the surface, we can speculate about the rate of movement along the pipe and the simultaneous rate of polymerization, i.e., about the diagonal path that the material will follow across the surface portrayed. As elsewhere, the diagram does not in itself create new information, but it helps us to visualize the simultaneous processes and their influence on each other; it could be a guide as to what factors to introduce if we wanted to set up a calculation for, for example, the rate of production of heat at different points along the pipe.

This concludes our initial survey of chemical-potential gradients and their influence on a material's responses. We now turn our backs on chemistry and consider some ideas of a strictly mechanical kind for a couple of chapters, in preparation for weaving the two trains of ideas together (see Figure 5.3).

6

NONHYDROSTATIC STRESS

As in the chapters on chemical potential, it will again be assumed that the reader has thought about the topic before, so that our task is to select rather than to build.

The interior of a continuous sample contains many small volumes and small areas, on any of which attention can be focused. A small internal area has the property that, across it, the material on one side exerts a normal force and a tangential force on the material on the other side. Let the normal force be F and the area A; then the ratio F/A approaches a limit as the size of A approaches zero. Thus we define the magnitude of the normal stress at a point across an infinitesimal area of a particular orientation. If we set up Cartesian coordinates so that the orientation of the area can be specified by the direction of its normal then, at a point, for every direction vector there is a normal-stress magnitude. The stress may be compressive or tensile, and in this text we treat compressions as positive.

Principal directions and stresses

It is possible to imagine a universe where space itself has an attribute of left-handedness or right-handedness, or where space does not but materials do. But if we set these possibilities aside and use ordinary ideas about symmetry, it follows that at any point where stresses exist inside a continuum, there are three orthogonal planes across which the tangential stress is zero; these planes suffer *only* normal stresses. The planes themselves are *principal planes*, their normals are the three *principal directions* at the point and the normal-stress magnitudes are the *principal stress magnitudes*. (See Figure 6.1.)

The largest, intermediate, and smallest normal compressions will be designated σ_1, σ_2, and σ_3, respectively; for most of what follows we shall designate the directions along which these compressions act as x_1, x_2, and x_3 (so that the plane compressed by stress σ_1 has x_1 for its normal), and we shall use x_1, x_2, and x_3 as axes for a local Cartesian system with which other planes and directions at the point can be specified. In particular, for any direction through the point, a *unit vector* can be imagined (magnitude $= 1$ unit of length); its components along the three axes will be called n_1, n_2, and n_3, combining to give the unit vector **n**. Another attribute of **n** is the set of three angles α, β, and γ as in Figure 6.2. We see that

$$n_1 = \cos \alpha, \qquad n_2 = \cos \beta, \qquad n_3 = \cos \gamma$$

40

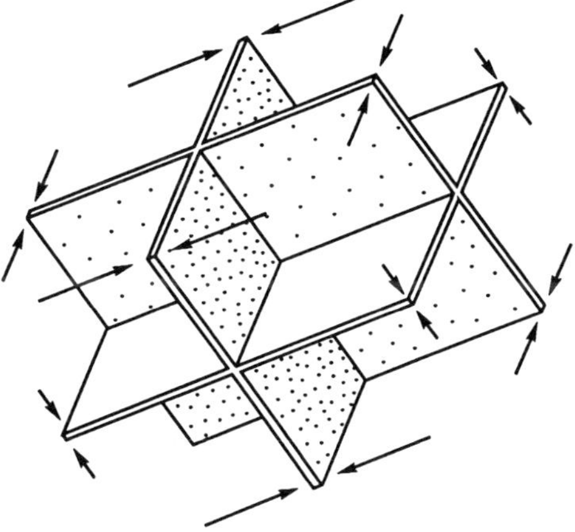

Figure 6.1 Three orthogonal planes that carry normal stresses but no shear stresses.

that is, n_1, n_2, and n_3 are the *direction cosines* of the direction **n** (sometimes alternatively labeled *l*, *m*, and *n*—*n* being the most overworked letter in the alphabet):

$$n_1^2 + n_2^2 + n_3^2 = 1 \qquad \text{or} \qquad l^2 + m^2 + n^2 = 1$$

This brings us to a fact that plays a central role in what follows: the compressive stress across a plane with normal **n** is $n_1^2\sigma_1 + n_2^2\sigma_2 + n_3^2\sigma_3$.

For an immediate example, let the principal stresses be 6, 4, and 3 MPa and let **n** be the direction that makes the same angle with all three axes so

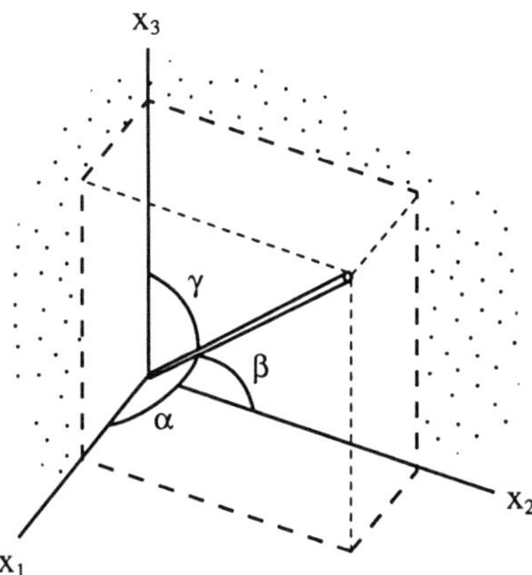

Figure 6.2 The angles between a unit vector and three reference directions.

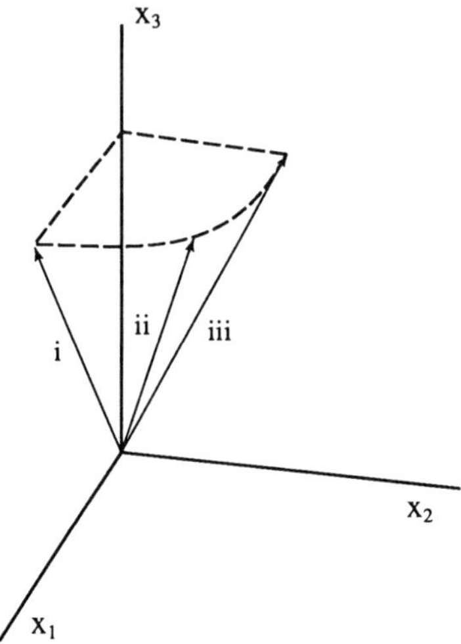

Figure 6.3 Three directions all making an angle α with direction x_3, where $\sin^2 \alpha = 0.2$ and $\cos^2 \alpha = 0.8$.

that $n_1^2 = n_2^2 = n_3^2 = 1/3$: then the compressive stress on the plane normal to **n** is $6/3 + 4/3 + 3/3 = 4\frac{1}{3}$.

For a slightly less special example, let us consider some directions **n** all close to the x_3 axis. Specifically let $n_3^2 = 0.8$; then the directions considered make an angle of about $26°$ with x_3, as in Figure 6.3. For the three directions in the figure, we have

(i) $n_1^2 = 0.2,$ $n_2^2 = 0$

(ii) $n_1^2 = 0.1,$ $n_2^2 = 0.1$

(iii) $n_1^2 = 0,$ $n_2^2 = 0.2$

Then compressive stresses on planes normal to the three directions are 3.6, 3.4, and 3.2 MPa, respectively. As we move away from the x_3 direction where the normal stress component is 3.0, the normal stress increases more rapidly for $26°$ of rotation toward x_1 and less rapidly for $26°$ of rotation toward x_2. Along the intermediate direction, the increase is interesting: in the first $26°$ of rotation the increase is only 0.4, whereas in rotating on to the direction considered first ($n_1^2 = n_2^2 = n_3^2 = 1/3$) the further increase is 0.93. The extra rotation is $29°$ so that the increase per degree of rotation is much larger.

The point just made is general and is used later: for directions close to a principal direction, the normal stress changes only slightly for say $5°$ change in direction, whereas for directions close to the octahedral direction the change in stress for a $5°$ change in direction is greater.

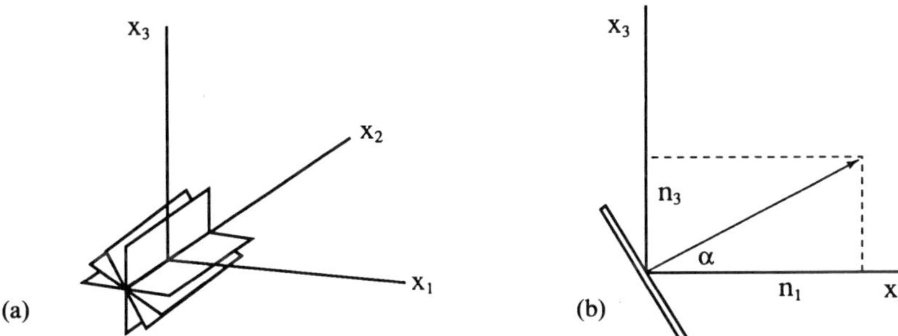

Figure 6.4 A set of planes whose normals all lie in the x_1x_3 plane, and the angle α that identifies one plane from this set.

Stress relations in the principal planes

Let us consider the set of planes in Figure 6.4, all parallel to direction x_2. For all of these, $n_2 = 0$ and only n_1 and n_3 are variables; the whole situation can be represented in two dimensions (6.4b) and

$$\sigma_n = n_1^2\sigma_1 + n_3^2\sigma_3 \tag{6.1a}$$

$$= (\cos^2 \alpha)\sigma_1 + (\cos^2 \gamma)\sigma_3 \tag{6.1b}$$

Also, in this special situation $\alpha + \gamma = 90°$ and $\cos \gamma = \sin \alpha$, so that

$$\sigma_n = \sigma_1 \cos^2 \alpha + \sigma_3 \sin^2 \alpha \tag{6.1c}$$

$$= \tfrac{1}{2}(\sigma_1 + \sigma_3) + \tfrac{1}{2}(\sigma_1 - \sigma_3) \cos 2\alpha \tag{6.1d}$$

See Figure 6.5, where diagram (b) is a Mohr diagram and diagram (a) will be called a *sine-wave diagram*.

Similar diagrams can be drawn for the 1,2 plane and the 2,3 plane, and the diagrams obviously have to fit together—see Figure 6.6. Also something not quite so obvious: if in Figure 6.6a we focus on the 1,3 "fence" and see a sine-wave bounded by the 1 level above and the 3 level below, and then transfer to Figure 6.7, where the levels are shown by spherical caps of radius σ_1 and σ_3, the sine-wave in Figure 6.6 reappears as an ellipse-portion in Figure 6.7.

Summary. The variation of compressive stress with direction in a principal plane can be represented in three equivalent ways—by an ellipse, a sine-wave or a Mohr circle. The variation is fixed by just the two extreme values, σ_1 and σ_3 or σ_2 and σ_3 or σ_1 and σ_2. The different diagrams are simply visualizations of the original statement

$$\text{Normal stress component} = n_1^2\sigma_1 + n_2^2\sigma_2 + n_3^2\sigma_3$$

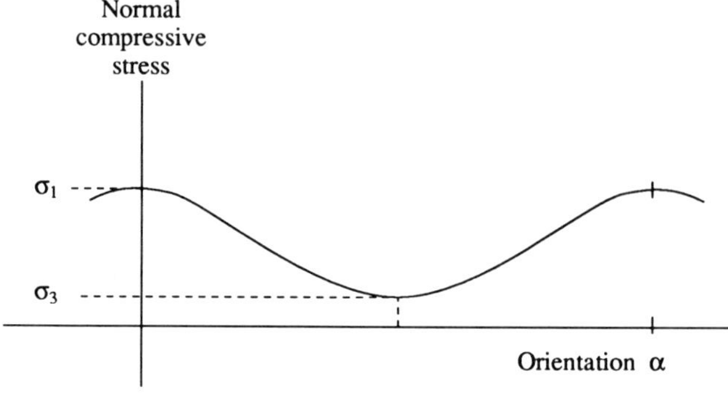

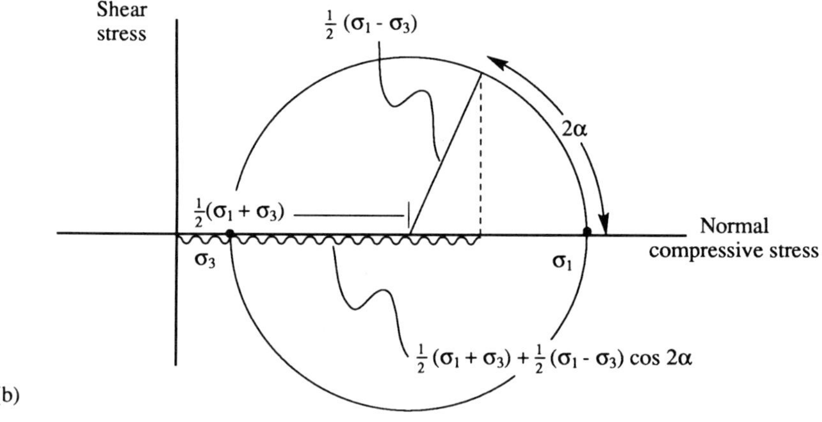

Figure 6.5 The normal stresses on a set of planes as shown in Figure 6.4a. (a) Variation of normal stress with orientation α. (b) The same range of normal stresses shown by a Mohr circle.

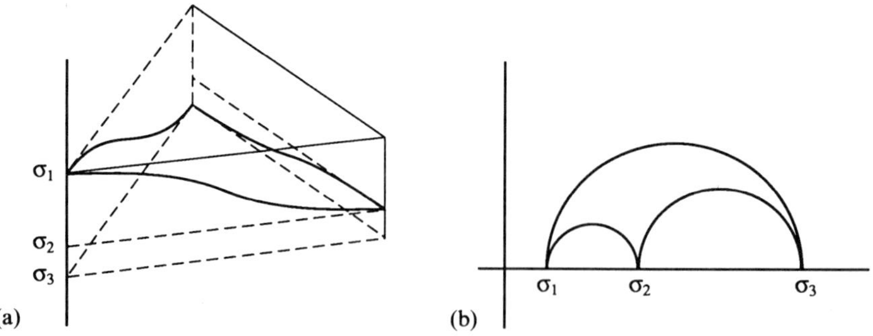

Figure 6.6 Normal stresses on three sets of planes. Each set of planes is of the type shown in Figure 6.4a; their normals lie in the x_2x_3 plane, the x_3x_1 plane, and the x_1x_2 plane, respectively.

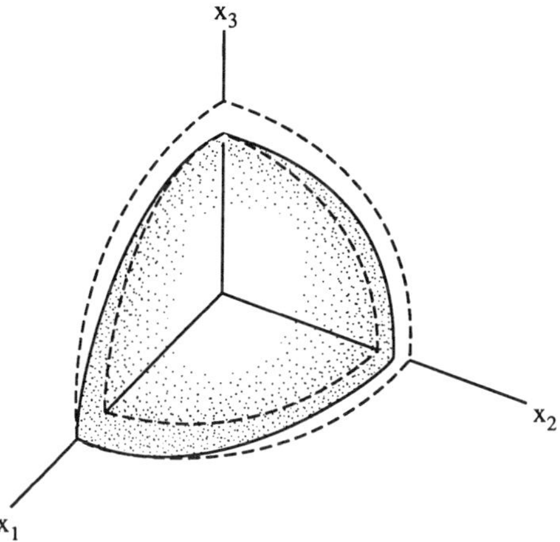

Figure 6.7 A surface drawn so that, for any direction **n**, the distance from the origin to the surface along direction **n** is proportional to the normal stress suffered by a plane whose normal is **n**. The surface lies wholly within the space between a sphere of radius σ_3 and a sphere of radius σ_1; these spheres are indicated by broken lines.

Tensile stresses

In some materials it is possible for an interior stress to be tensile. All the preceding algebraic relations persist and most of the diagrams persist with their characteristics unchanged: only the representation by three ellipses and an ellipsoid no longer works out.

Shear stresses

On any principal plane at any point inside a material, the shear stress is zero, but on every nonprincipal plane the shear stress has some nonzero magnitude—a material is normally full of shear stresses. But all their magnitudes are known functions of the principal stresses σ_1, σ_2, and σ_3; specifying these three specifies the stress state *completely*. Hence any behavior that could be seen as "caused by" a shear stress can be seen alternatively as "caused by" the stress trio σ_1, σ_2, σ_3.

The same point can be made in a different way as follows. Let us return to the plane shown in Figure 6.4b, whose normal lies in the x_1,x_3 plane and makes angle α with x_1. The shear stress on this plane, σ_t, is given by

$$\sigma_t = \tfrac{1}{2}(\sigma_1 - \sigma_3)\sin 2\alpha \qquad (6.2)$$

and we see at once that

$$\sigma_t = -\frac{1}{2}\frac{d}{d\alpha}(\sigma_n) \qquad (6.3)$$

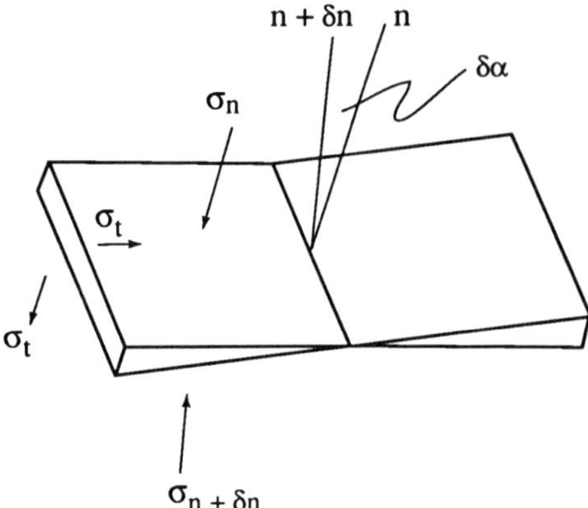

Figure 6.8 A wedge-shaped element as imagined in the interior of an extensive sample of stressed material. The stresses shown are associated with forces on the element; the components of these forces along a direction such as $\mathbf{n} + \delta\mathbf{n}$ must balance when the dimensions of the element approach zero.

Thus any behavior that could be seen as "caused by" the shear stress can be seen alternatively as "caused by" the fact that σ_n changes with α, or by the *gradient* $d\sigma_n/d\alpha$ that shows up so prominently in Figure 6.5a.

Before leaving the topic, we should generalize the result (6.3). Figure 6.8 shows an infinitesimal volume-element and some of the stresses acting on it. Balancing forces or conserving momentum shows that eqn. (6.3) must be correct, regardless of whether the normal $\mathbf{n}$ and the angle α lie in a principal plane or not. In other words, referring to Figure 6.6a, any gradient on the surface shown corresponds to a shear stress in the stress state represented. Where the gradient goes to zero, the shear stress goes to zero and where the gradient is a maximum the shear stress is a maximum. The link between shear stress and a normal-stress *gradient* is essential for succeeding chapters.

7
CHANGE OF SHAPE AND CHANGE OF VOLUME

In earlier chapters we first defined a material's chemical potential, and then went on to enquire how the material responds. And similarly with a state of nonhydrostatic stress: having reviewed what it is, we consider how a material might respond.

For the sake of simplicity, we imagine an extensive sample, such as a cubic meter, and suppose that the stress state is the same in every cubic centimeter; that is to say, there are no gradients in stress *from point to point*. Thus we do not enquire yet how a material responds to a spatial stress gradient; that comes later. We first enquire how it responds to a homogeneous but nonhydrostatic stress.

Strain and strain rate

Inside the material, close to the point of interest, we define a small length l by means of the material particles at its two ends. If, at a later moment, we find the distance between the particles to be $l - \delta l$, then we envisage the limit of the ratio $\delta l/l$ as l goes to zero, give the limit the symbol ε, and name it the *linear strain* at the point of interest in the direction of l, positive when δl is positive, i.e., for a shortening and negative for an elongation.

Another mental operation that can be performed in the neighborhood of the point of interest is to define a small sphere by means of the material particles that form its surface. At a later moment the particles will form the surface of an ellipsoid. (For a large sphere and an inhomogeneous situation, the new shape can be something more complicated; but as the imagined original sphere approaches zero diameter, the shape of its deformed counterpart can only approach an ellipsoid). The axes of the ellipsoid are principal directions of strain, and the magnitudes of the strains along them are named ε_1, ε_2, and ε_3, with ε_1 the largest. In an isotropic material, the principal axes of stress and strain coincide, with ε_1 lying along the direction of σ_1 and correspondingly; see Figure 7.1a.

As with stresses, the three values of ε themselves define an ellipsoid if they are all positive—see Figure 7.1b. The ellipsoid with diameters proportional to ε is a kind of inverse of the material ellipsoid formed from the initial material sphere: where the latter is most flattened (along the x_1 axis) ε takes its largest value and the ellipsoid of ε-values is most elongated. Figure 7.1

47

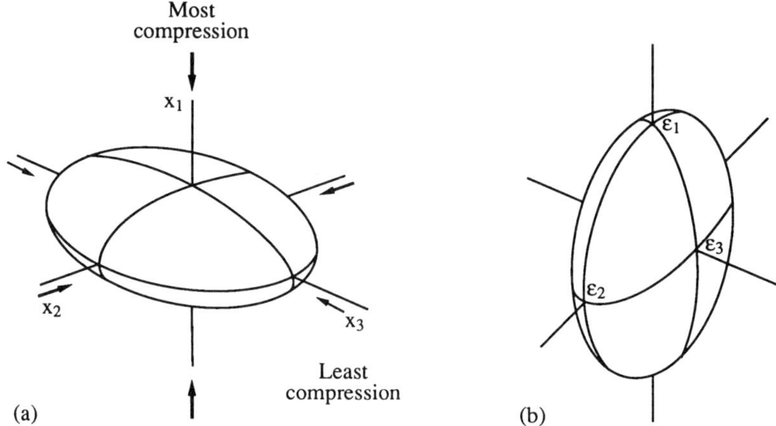

Figure 7.1 Homogeneous deformation of a sphere. If an original sphere is deformed by unequal compressions, with the compression along x_1 the greatest, the sphere's shape at a later moment might be as shown in (a). A more abstract representation of the change of shape is given in (b). Here the length of the radius drawn along axis 1 is proportional to the amount of shortening that occurs along axis 1, and correspondingly for all other directions through the center of the original sphere.

is deceptive, however; we are mostly concerned with changes where the material ellipsoid stays very close to a sphere. For purposes of illustration the departure from sphericity is greatly exaggerated in Figure 7.1a, whereas there is no exaggeration in Figure 7.1b.

Again in parallel with stresses, a general relation is

$$\varepsilon_n = n_1^2 \varepsilon_1 + n_2^2 \varepsilon_2 + n_3^2 \varepsilon_3 \tag{7.1}$$

and this gives rise to special relations in the symmetry planes:

$$\varepsilon_n = n_1^2 \varepsilon_1 + n_3^2 \varepsilon_3 \tag{7.2a}$$

$$= \varepsilon_1 \cos^2 \alpha + \varepsilon_3 \sin^2 \alpha \tag{7.2b}$$

$$= \tfrac{1}{2}(\varepsilon_1 + \varepsilon_3) + \tfrac{1}{2}(\varepsilon_1 - \varepsilon_3) \cos 2\alpha \tag{7.2c}$$

and diagrams resembling Figures 6.5 and 6.6 can be drawn, even when some or all of the principal strains are negative. An important distinction from the stress relations needs to be made, however: eqs. (7.1) and (7.2) become exact only as the strain magnitudes become small. (There are some strain product terms such as $\varepsilon_1 \varepsilon_2$: when the strains are small, these can legitimately be ignored, but for large strains they have to be written in.)

Strain rates. If we return to the definition of strain, and let the change δl occur during an interval of time δt, then again we can form a ratio, $\delta l/(l\,\delta t)$; we seek the limit of this ratio as δt goes to zero while l remains at some suitable small value. The strain rate is designated e; eqs. (7.1) and (7.2) can

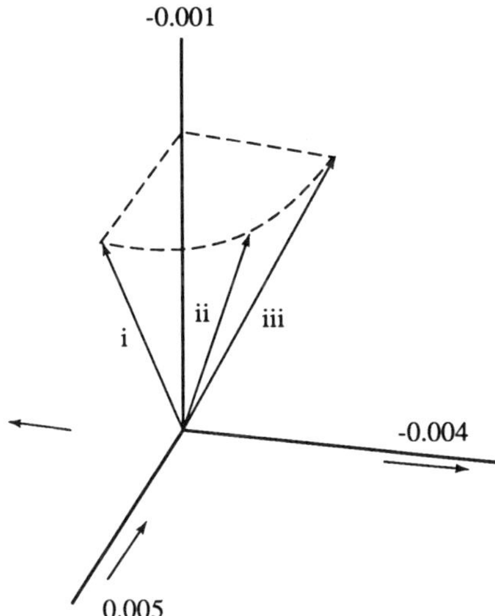

Figure 7.2 Amounts of shortening along three axial directions, and directions (i), (ii), and (iii) along which the shortening can be calculated.

be rewritten with e for ε throughout; and because we are now taking a limit as δl goes to zero while l does not, the equations as written are complete and exact, as long as they are relations among strain *rates*.

For some numerical examples we can consider the same directions as in Figure 6.3. Let the shortening rate be 0.005 per day along x_1, with elongation rates 0.004 per day along x_2 and 0.001 per day along x_3, as in Figure 7.2. Then along direction (i) the shortening rate is $(0.8)(-0.001) + (0.2)(0.005) = 0.0002$, and along directions (ii) and (iii) there are elongation rates 0.0007 and 0.0016. In this example, if we consider the octahedral direction $n_1^2 = n_2^2 = n_3^2 = \frac{1}{3}$, we find that the strain rate is exactly zero.

Change of shape and change of volume

For another numerical example, let the strains or strain rates per day be 6, 4, and 3 thousandths, all shortenings. A useful operation is to treat this change as the sum of two changes superimposed, viz. $4\frac{1}{3}$, $4\frac{1}{3}$, $4\frac{1}{3}$ *plus* $1\frac{2}{3}$, $-\frac{1}{3}$, $-1\frac{1}{3}$. The usefulness results from the fact that the first involves no change of *shape* (whatever shape one had in view initially, the result is the same shape but just slightly smaller) and the second involves no change in *volume*. Here $4\frac{1}{3} = \frac{1}{3}(6 + 4 + 3)$, and in general the most useful partition of e_1, e_2, e_3 is into e_0, e_0, e_0 *plus* $e_1 - e_0$, $e_2 - e_0$, $e_3 - e_0$ where $e_0 = (e_1 + e_2 + e_3)/3$. And why is this partition useful? Let us partition the stress state the same way, into $\sigma_0 = (\sigma_1 + \sigma_2 + \sigma_3)/3$ and $\sigma_1 - \sigma_0$, $\sigma_2 - \sigma_0$, $\sigma_3 - \sigma_0$. Then we can see that the usefulness is that σ_0/e_0 is one property of the material—its ability to resist change of *volume*—while $(\sigma_1 - \sigma_0)/(e_1 - e_0)$ is a different property—its ability to resist change of *shape*. The difference between these is illustrated by a rubber bag stuffed with ball-bearings: it is reasonably easy to change the bag's shape, but very difficult to change its volume (except to the slight extent that goes with the bearings being ideally or not so ideally close-packed).

A point to consider quickly is the matter of the three ratios $(\sigma_1 - \sigma_0)/(e_1 - e_0)$, $(\sigma_2 - \sigma_0)/(e_2 - e_0)$, and $(\sigma_3 - \sigma_0)/(e_3 - e_0)$. For many real materials these are very close to equal and we shall consider only ideal materials where they are exactly equal. The abbreviations $e_i' = e_i - e_0$ and $\sigma_i' = \sigma_i - \sigma_0$ are useful ($i = 1$, 2, or 3). If σ_i'/e_i' has the same value C for all i, then for any direction

$$\frac{\sigma_\mathrm{n}'}{e_\mathrm{n}'} = C \tag{7.3}$$

where $\mathbf{n}$ is the unit vector in that direction. To see the validity, compare eqn. (7.1) with the corresponding expression for σ_n. In description of ideal elastic materials, where strains are small, instantaneous, and totally recoverable, symbols often used are

$$\frac{\sigma_0}{\varepsilon_0} = 3K \qquad (K = \text{bulk modulus}) \tag{7.4}$$

and

$$\frac{\sigma'}{\varepsilon'} = 2G \qquad (G = \text{shear modulus}) \tag{7.5}$$

Another pair of properties serving the same purpose as K and G is the pair λ and μ, where $\mu = G$ and $\lambda = K - \frac{2}{3}G$; for keeping change of shape separate from change of volume, the pair K and G are preferred.

In description of ideal viscous materials, where strains are totally non-recoverable,

$$\frac{\sigma_0}{e_0} = \text{volume viscosity or second viscosity coefficient} \tag{7.6}$$

and

$$\frac{\sigma'}{e'} = 2N \qquad (N = \text{viscosity}) \tag{7.7}$$

The second viscosity coefficient is often assumed to be infinite and in any case requires further discussion, but K, G, and N are straightforward material properties. The separation of change of volume from change of shape is made even more complete if we write, as consequences of eqns. (7.5) and (7.7),

$$\frac{\sigma_1 - \sigma_3}{\varepsilon_1 - \varepsilon_3} = 2G \qquad \frac{\sigma_1 - \sigma_3}{e_1 - e_3} = 2N \qquad (7.8\text{a,b})$$

and similarly for other pairs of principal directions. Here the mean values σ_0 and e_0 are not involved, even implicitly.

Units and examples. Because the strain ε is just a ratio (dimensionless), the units of the shear modulus G are the same as those of the stress difference $\sigma_1 - \sigma_3$, such as megapascals, bars, or pounds per square inch (psi):

$$1 \text{ MPa} = 10 \text{ bar} \cong 145 \text{ psi} \cong 10 \text{ kg/cm}^2$$

Suppose a person stands on a substantial block of firm cheese and the height of the block diminishes by 1 mm. First let us assume the effect is purely elastic at constant volume and consider what the material's shear modulus might be. Assume height of block $= 160$ mm. Then strain $\varepsilon_1 = 1/160$ taking vertical as direction 1. Assume horizontal strains ε_2 and ε_3 are equal; then for constant volume, $\varepsilon_2 = \varepsilon_3 = -1/320$ and $\varepsilon_1 - \varepsilon_3 = 3/320$.

For the magnitude of σ_1, assume 72 kg on 120 cm^2, giving 0.6 bar or 0.06 MPa. Also in this test, $\sigma_2 = \sigma_3 = 0$. Hence $2G = (0.06 \text{ MPa}) \times 320/3$ or 6.4 MPa.

On the other hand, suppose the 1 mm shortening were a purely viscous effect that required 10 seconds of applied stress: what would be the viscosity of the material? The driving stress difference is still 0.06 MPa, but the denominator needed (eqn. 7.8b) is now a strain *rate* 3/3200 per sec. Hence $2N = 64$ MPa-sec or 3840 MPa-min or 64×10^6 Pa-sec or 64×10^7 poise.

$$1 \text{ MPa-sec} = 10 \text{ bar-sec} = 10^6 \text{ Pa-sec} = 10^7 \text{ poise}$$

To consolidate the ideas, let us calculate a few more details: first the mean stress $(\sigma_1 + \sigma_2 + \sigma_3)/3 = 0.02$ MPa; then $\sigma_1' = 0.04$ and $\sigma_2' = \sigma_3' = -0.02$ MPa. Then by eqn. (7.7), $e_1' = 0.04/64$ or 2/3200 per sec and $e_2' = -1/3200$ per sec. These are the same as the rates e_1 and e_2, respectively; this is as it should be because it was assumed at the outset that the deformation occurred at constant volume, in which case $e_0 = 0$ and $e_i = e_i'$.

For an example where the volume changes, let us go back to an elastic material and use the material properties:

$$3K = 12 \text{ MPa}, \qquad 2G = 10 \text{ MPa}$$

but the same stress state as before, $\sigma_1 = 0.06$ MPa, $\sigma_2 = \sigma_3 = 0$, $\sigma_0 = 0.02$ MPa. Then by eqn. (7.4), $\varepsilon_0 = 1/600$ and by (7.5) $\varepsilon_1' = 0.04/10 = 1/250$ and $\varepsilon_2' = -1/500$. The difference $\varepsilon_1' - \varepsilon_2' = \varepsilon_1 - \varepsilon_2 = 3/500$, so eqn. (7.8a) checks out. The changes of shape ε_1', etc. are smaller than in the constant-volume example because the value of $2G$ is now larger, but the total shortening strain ε_1 is almost as large as before, because ε_1 combines ε_1'

plus ε_0 ($1/600 + 1/250 = 0.0057$ compared with $1/160 = 0.0063$ at constant volume).

Summary

The principal strain magnitudes at a point are a set of three numbers comparable with the principal stresses; diagrams resembling Figures 6.5–6.7 can be drawn and equations resembling eqn. (6.1) written (eqn. 7.2); but the strain equations are exact only for small strains.

The principal strain rates are a similar set again, and the strain-rate equations are exact.

It is convenient to separate σ_1, σ_2, σ_3 into σ_0, σ_0, σ_0 and σ_1', σ_2', σ_3', and similarly for ε_1, ε_2, ε_3. Then useful material properties can be defined: see eqns. (7.4–7.7).

Shear strains

It will be recalled that a stress state at a point is completely specified by the three magnitudes of the principal normal stresses; shear stresses certainly are present but their magnitudes are not additional independent quantities. Similarly with strains, shear strains exist, but they are details inside a state that is already fully specified by the three principal linear strains ε_1, ε_2, and ε_3.

To define a shear strain, we define two small, straight material lines inside a sample that meet at a point and at some moment form a right angle. At a later moment, let the angle they form differ from a right angle by θ; then the shear strain suffered by the pair of lines is $\frac{1}{2}\tan\theta$. If the directions of the two lines are named i and j the shear strain can be written ε_{ij}. If a small strain $\delta\varepsilon_{ij}$ occurs in a small period of time δt, the shear strain rate at some moment, e_{ij}, can be taken as the limit of $\delta\varepsilon_{ij}/\delta t$ as $t \to 0$, and

$$e_{ij} = \frac{1}{2}\frac{d}{d\theta}(e_{ii}) \qquad (7.9)$$

Here the angle θ necessarily lies in the ij plane so that a change in θ is a rotation about the k direction, and this fact is conveniently recorded by writing θ_k; e_{ii} is the linear strain rate in the i direction, and we do not use the summation convention.

From eqn. (7.7), $d(e_{ii})/d\theta_k = (1/2N)d(\sigma_{ii})/d\theta_k$, so that

$$e_{ij} = \frac{1}{4N}\frac{d(\sigma_{ii})}{d\theta_k} \qquad (7.10)$$

A point to be noted about eqn. (7.10) is its resemblance to other phenomenological "laws" where a flux is driven by a gradient, for example, Fick's first law,

Diffusive flow rate = (material property) × (concentration gradient)

Equation (7.7) does not have this form, nor does its tensor equivalent $\boldsymbol{\sigma}' = 2N\mathbf{e}'$. The overall intention of these chapters is to approach problems where Fickian diffusion and viscous change of shape interact, and the fact that eqn. (7.10) has roughly a Fickian form will be used again later.

In fact, eqn. (7.10) is something of a staging point or milestone. Up to this point, the purpose has been to review various basic ideas that we need, which are fully written up in other textbooks. But eqn. (7.10) is more novel; I have not seen it in any other book. The fact that the normal stress σ_{ii} has a derivative or gradient with respect to orientation θ is obvious in the sine-wave diagram, Figure 6.5a, but in other books the fact has been little used. By contrast, here we emphasize that fact and link it to the material's change of shape: the strain rate e_{ij} is an *effect* and the gradient $d\sigma/d\theta$ is its *cause* or, in conventional use of the words, e_{ij} is a *flux* and $d\sigma/d\theta$ is its accompanying *force*. It is this point of view that is new; one might say that the purpose of the book is to take this point of view and explore its consequences.

8

CONSERVATION

To show the purpose of this chapter, Figure 8.1 is repeated from the end of Chapter 5. The idea we use to draw together the two threads is the idea of *conservation*, e.g., of heat or mass.

Conservation statements

The central thought concerns a small volume inside a continuum, which we imagine together with its bounding surface: if we focus on something that can flow across the surface, such as heat, and designate by Q the quantity of heat inside the surface, then:

$$\frac{dQ}{dt} = (\text{rate of flow in}) - (\text{rate of flow out})$$

$$+ \text{ rate of creation of heat by processes occurring within the boundary}$$

Where mass is the thing that flows, one is often justified in assuming that no mass is created or destroyed within the boundary; when Q designates mass then,

$$\frac{dQ}{dt} = (\text{rate of flow in}) - (\text{rate of flow out})$$

A volume-element that is particularly simple to imagine is an orthogonal

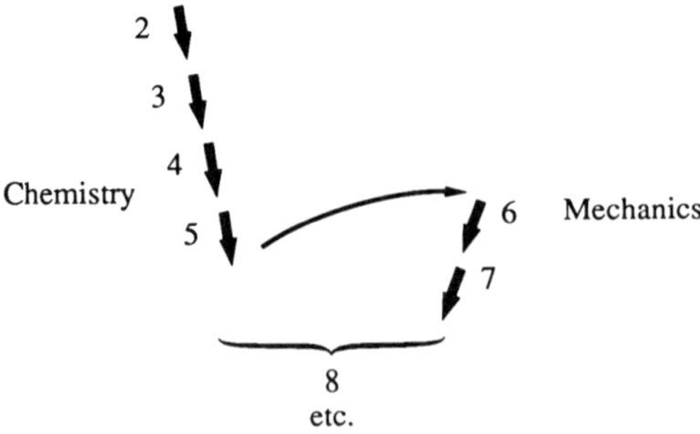

Figure 8.1 A flow chart showing the sequence of chapters (Figure 5.3 repeated).

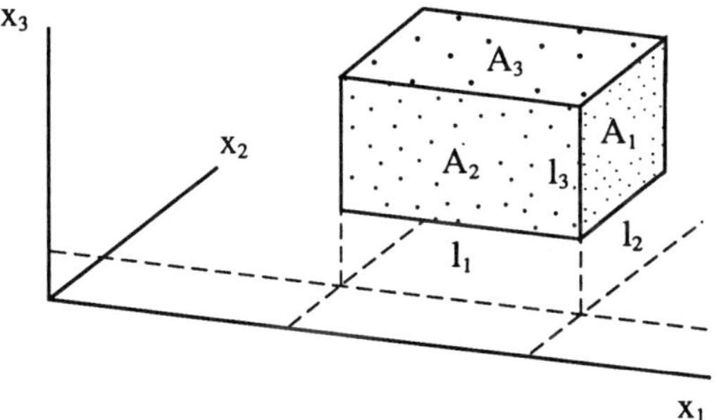

Figure 8.2 A volume-element through which flow occurs.

block as in Figure 8.2. For the pair of faces normal to x_1,

$$\text{Flow in at left} = A_1(\text{flux at left}) \tag{8.1a}$$

and

$$\text{Flow out at right} = A_1(\text{flux at right}) \tag{8.1b}$$

so

$$\left(\frac{dQ}{dt}\right)_1 = A_1(\text{difference between fluxes}) \tag{8.2a}$$

$$= A_1\left(\frac{\partial \text{ flux}}{\partial x_1}\right)l_1 \tag{8.2b}$$

$$= (\text{volume})\left(\frac{\partial \text{ flux}}{\partial x_1}\right) \tag{8.2c}$$

Also,

$$\text{Flux} = \text{conductivity} \times \text{driving gradient} \tag{8.3}$$

and

$$\text{Driving gradient} = \frac{\partial(\text{potential})}{\partial x_1} \tag{8.4}$$

so

$$\frac{(dQ/dt)_1}{\text{Volume}} = \frac{\partial}{\partial x_1}\left(\text{conductivity} \times \frac{\partial \text{ potential}}{\partial x_1}\right) \tag{8.5}$$

In the special case where conductivity does not vary with position in the sample, this gives

$$\frac{(dQ/dt)_1}{\text{Volume}} = K\frac{\partial^2 \text{ potential}}{\partial x_1^2} \tag{8.6}$$

and, summing over all three pairs of faces,

$$\frac{dQ/dt}{\text{Volume}} = K\left(\frac{\partial^2 \mu}{\partial x_1^2} + \frac{\partial^2 \mu}{\partial x_2^2} + \frac{\partial^2 \mu}{\partial x_3^2}\right) \tag{8.7}$$

if we write μ for the potential. To supplement this equation with a physical picture, we need to consider, first, what type of gradient might drive a flow of mass and, second, what type of material might flow or permit flow to occur.

Gradients. The gradients that come readily to mind are those of pressure and concentration. Also, of course, a temperature gradient drives flow of heat and has a slight tendency to drive flow of matter, and in addition there are gradients with respect to internal-variable coordinates. But the most conspicuous gradients driving flow of matter are those of concentration and pressure, each of which can be viewed as a gradient in a component's chemical potential (see Chapter 3).

Materials. For a concentration gradient to drive a flow, one has to have two components. Also, at least one must have a sufficient degree of geometrical continuity to serve as a reference-frame—because a concentration gradient does not drive a component through space: it only drives one component through the other component.

In a similar way, a pressure gradient may drive one of two components through or past the other. A fluid in a permeable host, such as water percolating through a dike of sand, is an example. Here we distinguish two possibilities: either the permeable host can change porosity or its porosity is fixed. In the former case, if flow into a volume-element exceeds flow out, the porosity can increase—the sand can swell to accommodate the extra fluid. In the latter case, of fixed porosity, the flow of one fluid can occur only to the extent that some second fluid permits it. Examples are an open porous brick of sandstone, where, as water flows in, air can flow out; or a porous oil-saturated sandstone where as water flows in, oil flows out. The fact that the equations are rather similar whether the host is deformable or not is illustrated by Figure 8.3.

In Figure 8.3, (a) shows a fluid in a deformable host (stippled); initially the host is of uniform thickness, with a uniform content of fluid, and flow can only occur to left or right in the plane of the diagram (no flow perpendicular to the page). The host is instantaneously loaded with a set of heavy rods whose lengths form a sine curve; then the pressure in the fluid-filled layer also must vary sinusoidally along x:

$$P = P_0 + j \sin x, \qquad \frac{dP}{dx} = -j \cos x \tag{8.8a,b}$$

$$\frac{dQ/dt}{\text{volume}} = K\frac{d^2 P}{dx^2} = Kj\frac{d^2(\sin x)}{dx^2} \tag{8.9a,b}$$

$$= -Kj \sin x \tag{8.9c}$$

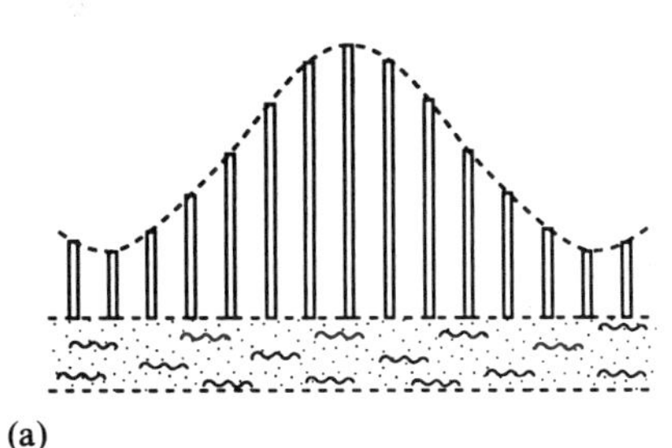 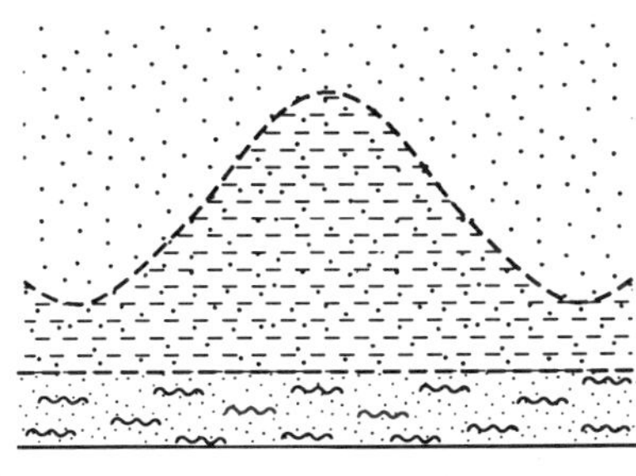

(a)   (b)

Figure 8.3 A layer of fluid-filled permeable material that is subjected to a spatially nonuniform load. The layer is seen edge-on at the bottom of each picture and is imagined to extend indefinitely perpendicular to the page. In (a) we imagine the load to be imposed by heavy rods of unequal length; in (b), we imagine a permeable material overlying the layer of interest, and a second fluid occupying the pore space of this material to nonuniform depths.

The host diminishes in volume most rapidly under the peak and swells most rapidly under the troughs; halfway down the flanks, the host neither shrinks nor swells, though this is the region where through-flow of fluid is the strongest.

In Figure 8.3b, the host material is more extensive and is assumed to be non-deformable. Again, in the starting state there is a layer of uniform thickness where the host is saturated with water; again, instantaneously, a load is applied—this time, by filling a sine-wave-shaped portion of the host above the water with oil. Then the pressure in the layer of water will again vary sinusoidally according to eqn. (8.8) and the flow of water will go according to eqn. (8.9). The only difference is that as water flows away to left and right from the region under the peak, the load/water interface descends through the rigid host, whereas in diagram (a), the host moved as the load/water interface moved.

Of course, in both these examples, after the first moment, complications would set in: the deformable host might begin to support some of the load, and in either case the layer of water would soon cease being a *layer of uniform thickness*. But none of these later complications affects the simplicity of the relations between the pressure distribution and the resulting behavior of Q at the instant that the pressure distribution first exists. We must keep attention fixed on this first moment of geometrical simplicity while extending the preceding ideas to two more situations.

Self-diffusion

We have to recognize that self-diffusion really occurs: materials are in fact atomic and a material can diffuse through itself. To pin the ideas down, one imagines a small interval of time: then, for that period of time, the atoms in a sample of material can be divided into those that change neighbors during the interval by making a jump and those that do not. These two groups of atoms play the roles of fluid and host (water and sand, or whatever they may be).

Usually self-diffusion is thought of as being affected by temperature—the hotter the material, the more active the atoms—and not thought of as being affected by pressure gradients; the diffusion is admitted to go on, but in a symmetrical way, with as many atoms jumping to the left as jump to the right. But the analogy with water and sand prompts the speculation: in a pressure gradient, will self-diffusion have a one-directional trend? Referring to Figure 8.3a, it is proposed that, even when the basal layer is a single pure material, the two processes *deformation of host* and *movement of fluid through host* can be distinguished; the distinction is made by the device used above of considering a short interval of time and dividing the atoms into two groups. During a short period, most atoms stay bonded as they were and act as a frame, but a population of diffusers acts like a fluid and migrates from high-pressure sites toward low-pressure sites. The effect is both inconspicuous and small but must exist: it is *different from* and *additional to* all those processes one may visualize that occur when a material deforms in a stress field that is homogeneous through space.

Of course one has to add that every atom gets a turn at being a diffuser; an atom that is a diffuser in one interval of time is part of the host during many succeeding intervals, and equally the host-material has no long-term permanence—every atom has the potential sooner or later to become mobile and diffuse away. One reaches a strange conclusion, that all the material moves by moving through itself; that seems contradictory, like lifting oneself by one's bootstraps, but it is not contradictory—the device of separating time into small intervals enables us to focus on some genuine effects in the real world.

Conservation and strain

We have already remarked that the shear–strain-rate equation (7.10) is a flux/driving-gradient relation, of the same type as eqn. (8.3). If a first-derivative equation (8.3) plus a conservation statement leads to second-derivative equations (8.6) and (8.7), we can enquire whether eqn. (7.10) has consequences of a parallel kind.

For an answer, recall that for any direction i, a normal stress σ_i' was defined by $\sigma_i' = \sigma_i - \sigma_0$, where σ_i is the full normal compressive stress acting on a plane normal to i, and σ_0 is the mean stress $(\sigma_1 + \sigma_2 + \sigma_3)/3$. It is shown in Appendix 8A that

$$\sigma_i' \quad \text{or} \quad \sigma_i - \sigma_0 = -\frac{1}{6}\left(\frac{\partial^2 \sigma_i}{\partial \theta_1^2} + \frac{\partial^2 \sigma_i}{\partial \theta_2^2}\right) \tag{8.10}$$

Here, as shown in Figure 8.4, $\delta\theta_1$ is a small angle away from the direction i in an arbitrary plane containing i, and $\delta\theta_2$ is a small angle away from

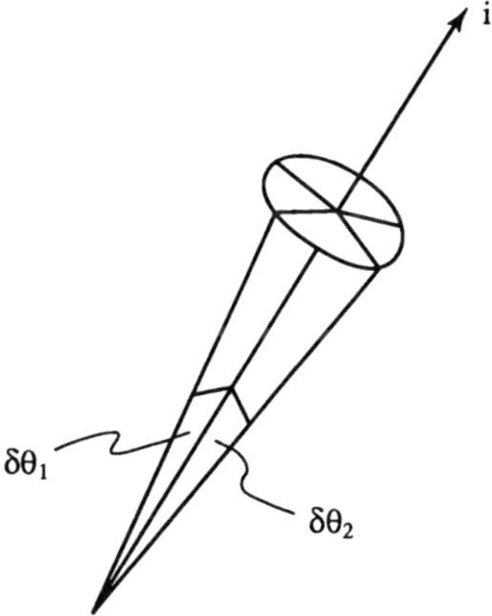

Figure 8.4 A line with direction i, two orthogonal planes that intersect in the line, and two small angles $\delta\theta_1$ and $\delta\theta_2$ measured from the line in the two planes.

direction i in a plane orthogonal to the first one. Then eqn. (7.7) immediately gives

$$e_i' = \frac{-1}{12N}\left(\frac{\partial^2\sigma_i}{\partial\theta_1^2} + \frac{\partial^2\sigma_i}{\partial\theta_2^2}\right) \tag{8.11a}$$

To see the correspondence with Fick's laws, etc., we repeat eqn. (7.10):

$$e_{ij} = \frac{1}{4N}\frac{\partial(\sigma_{ii})}{\partial\theta_k} \tag{7.10}$$

and rewrite eqn. (8.11a) in parallel notation:

$$e_{ii}' = \frac{-1}{12N}\left(\frac{\partial^2\sigma_{ii}}{\partial\theta_j^2} + \frac{\partial^2\sigma_{ii}}{\partial\theta_k^2}\right) \tag{8.11b}$$

Then we see that the shear strain rate e_{ij} corresponds to the flux through a plane in Fickian diffusion, and the linear shortening rate e_{ii}' corresponds to the gain or loss term dQ/dt per unit volume.

Summary

A diagram of essential relations is shown in Figure 8.5. Each curve represents an equation. The two plain arcs are the conventional equations for viscous behavior: both result from defining the viscosity N by the relation

(Radius of the Mohr stress circle)

$$= 2N \times \text{(radius of the Mohr strain-rate circle)}$$

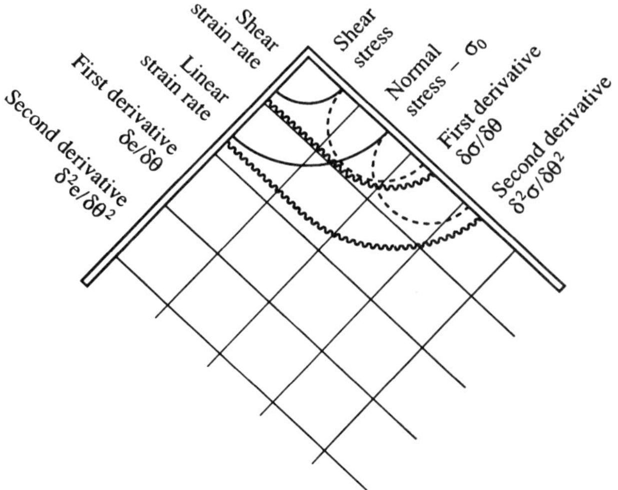

Figure 8.5 A diagram of relations among stresses and strain rates. The two full-line arcs are based on the definition of a material's viscosity, and the two dotted arcs are based on mechanical equilibrium or conservation of momentum. From these, two relations are derived that resemble Fick's first and second laws, represented by wiggly lines.

The upper dotted arc relates shear stress to $\partial(\text{normal stress})/\partial\theta$; it derives from the thin-wedge diagram, Figure 6.8, by equilibrium of forces or conservation of momentum. The lower dotted arc is also at bottom a mechanical equilibrium or conservation-of-momentum statement. It is by combining these four that we gain the two wiggly arcs, the two new equations in which a gradient of normal stress with respect to orientation is seen as the gradient that drives the material's change-of-shape response.

The powerful idea already mentioned in earlier chapters is that it is a gradient in *potential* that drives a material's response. Thus we come to a key question: can there be a gradient in *potential* with respect to orientation at a single point in space? This is the topic of Chapter 9.

APPENDIX 8A: PROOF OF THE STRESS RELATION (8.10)

For any plane P through the point of interest, we can consider the set of planes whose normals lie in plane P. For this set of planes, the normal stress has a maximum σ_z and minimum σ_x. Then for any other plane in the set with unit normal **n**,

$$\text{Normal stress}\quad \sigma_n = \bar{\sigma} + \frac{\sigma_z - \sigma_x}{2}\cos 2\theta \quad \left(\bar{\sigma} = \frac{\sigma_z + \sigma_x}{2}\right)$$

also

$$\frac{d\sigma_n}{d\theta} = -(\sigma_z - \sigma_x)\sin 2\theta$$

and

$$\frac{d^2\sigma_n}{d\theta^2} = -2(\sigma_z - \sigma_x)\cos 2\theta$$

so

$$\sigma_n = \bar{\sigma} - \frac{1}{4}\frac{d^2\sigma_n}{d\theta^2}$$

To keep track of the fact that $\bar{\sigma} = (\sigma_{max} + \sigma_{min})/2$ for plane P and that $d\theta$ lies in plane P, we write this as

$$\sigma_n = \bar{\sigma}_P - \frac{1}{4}\frac{d^2\sigma_n}{d\theta_P^2}$$

For a second plane Q that also contains direction **n**,

$$\sigma_n = \bar{\sigma}_Q - \frac{1}{4}\frac{d^2\sigma_n}{d\theta_Q^2}$$

so that also

$$\sigma_n = \frac{\bar{\sigma}_P + \bar{\sigma}_Q}{2} - \frac{1}{8}\left(\frac{\partial^2\sigma_n}{\partial\theta_P^2} + \frac{\partial^2\sigma_n}{\partial\theta_Q^2}\right) \tag{8A.1}$$

We now have to evaluate $(\bar{\sigma}_P + \bar{\sigma}_Q)/2$: consider three orthogonal planes P, Q, and R with lines of intersection l, m, n (not direction cosines). By the argument used above,

$$\sigma_l = \frac{\bar{\sigma}_Q + \bar{\sigma}_R}{2} - \frac{1}{8}\left(\frac{\partial^2\sigma_l}{\partial\theta_Q^2} + \frac{\partial^2\sigma_l}{\partial\theta_R^2}\right)$$

$$\sigma_m = \frac{\bar{\sigma}_R + \bar{\sigma}_P}{2} - \frac{1}{8}\left(\frac{\partial^2\sigma_m}{\partial\theta_R^2} + \frac{\partial^2\sigma_m}{\partial\theta_P^2}\right)$$

$$\sigma_n = \frac{\bar{\sigma}_P + \bar{\sigma}_Q}{2} - \frac{1}{8}\left(\frac{\partial^2\sigma_n}{\partial\theta_P^2} + \frac{\partial^2\sigma_n}{\partial\theta_Q^2}\right)$$

Summing:

    (i) $3\sigma_0 = \bar{\sigma}_P + \bar{\sigma}_Q + \bar{\sigma}_R$ (all the second derivatives cancel in pairs).

Also

    (ii) $3\sigma_0 = \sigma_n + \sigma_R^{max} + \sigma_R^{min} = \sigma_n + 2\bar{\sigma}_R$.

We eliminate $\bar{\sigma}_R$ by subtracting $\frac{1}{2} \times$ (eqn. ii) from eqn. (i):

$$\tfrac{3}{2}\sigma_0 = \bar{\sigma}_P + \bar{\sigma}_Q - \tfrac{1}{2}\sigma_n \qquad \text{or} \qquad \frac{\bar{\sigma}_P + \bar{\sigma}_Q}{2} = \tfrac{3}{4}\sigma_0 + \tfrac{1}{4}\sigma_n \qquad (8A.2)$$

Putting (8A.2) into (8A.1) gives

$$\sigma_n = \sigma_0 - \frac{1}{6}\left[\frac{\partial^2 \sigma_n}{\partial \theta_P^2} + \frac{\partial^2 \sigma_n}{\partial \theta_Q^2}\right]$$

APPENDIX 8B: CONSERVATION: A NUMERICAL EXAMPLE

The flow of heat provides a concrete instance. Suppose that in a long bar the temperature in K or °C varies according to

$$T = T_0 + 18 \sin 2\pi x/60$$

where x is measured in millimeters along the bar, and suppose the bar has cross-section 60 mm². Then a section of bar from one cool spot to the next has roughly the size of one's finger and is 36 K hotter in the middle than at the ends. If the material's conductivity were 3 μJ/mm-sec-K and we consider a slice 10 mm wide centered on the point $x = 15$ mm, a temperature maximum, at what rate would the slice be losing heat? We will calculate the answer by two methods and compare the results.

First consider the segment's two bounding surfaces, at 10 mm and 20 mm along x. The gradient at each, dT/dx, is $18(2\pi/60) \cos 2\pi 10/60 = 0.94$ K/mm, either positive or negative. Hence the flux at each is $3 \times 60(0.94)$ or 169 μJ/sec and, both fluxes being outward, the total rate of loss of heat is 338 μJ/sec.

Second, use eqn. (8.6) in the form

$$\frac{dQ}{dt} = K \frac{d^2(\text{potential})}{dx^2} \ (\text{segment's volume})$$

The second derivative is $18(2\pi/60)^2 \sin 2\pi 15/60 = 0.197$ K/mm², so that $dQ/dt = 3(0.197)600 = 355$ μJ/sec.

The two answers roughly agree, the first being the more accurate. The second is an overestimate because we evaluated the second derivative at $x = 15$ mm but then assumed that the value could be applied to the entire segment from 10 mm to 20 mm along x. To evaluate the second derivative at $x = 10$ mm would give an underestimate for the segment as a whole, but if we evaluate it at, say, $x = 12$ mm, the second derivative $= 0.187$ K/mm² and the rate of loss $= 337$ μJ/sec, in good agreement. The calculation serves as a helpful reminder than eqn. (8.6) applies exactly to an infinitesimal volume, so that applying it to a finite volume can only give an approximate result. Of course, in the main text, what we need and use is the exact infinitesimal form; it is only in the present numerical example that approximations enter.

APPENDIX 8C: INVARIANTS

The ideas presented in this appendix are not used in Chapters 9 through 18, and their use in Chapter 19 is only as an additional check or alternative point of view. But it is an essential check and a very useful point of view. Also, some parts of the two preceding chapters can be concisely reviewed by means of statements about invariants. Hence the concepts are introduced at this point.

As has been emphasized, the state of stress at a point can be expressed in three numbers, σ_{max}, σ_{min}, and σ_{int}, as long as we remember that the planes on which these act are orthogonal as in Figure 6.1 and that the whole stress state is symmetrical with respect to these planes. Given the three magnitudes, one can complete any of the representations in Figures 6.6 and 6.7. In particular, both parts of Figure 6.6 show the three magnitudes as points on a line. Any such group of three values clearly has a mean value, a degree of scatter or spread and a degree of lopsidedness—the last being zero when σ_2 is midway between σ_1 and σ_3, and being at an extreme when σ_2 equals σ_1 or σ_3. These three measures resemble the mean, the standard deviation and the kurtosis of a larger set of numbers; with only three magnitudes in view, the three measures completely specify the set as effectively as stating the individual values.

For a start, we define

$$\text{Mean stress } \sigma_0 = (\sigma_1 + \sigma_2 + \sigma_3)/3$$

$$\text{sum}_{II} = (\sigma_1 - \sigma_0)^2 + (\sigma_2 - \sigma_0)^2 + (\sigma_3 - \sigma_0)^2$$

$$\text{sum}_{III} = (\sigma_1 - \sigma_0)^3 + (\sigma_2 - \sigma_0)^3 + (\sigma_3 - \sigma_0)^3$$

Then, to have all measures in the same units, we take $(\text{sum}_{II})^{1/2}$ and $(\text{sum}_{III})^{1/3}$. To emphasize the difference between the invariants and the principal stresses, we use the symbols S_I, S_{II}, and S_{III}. Thus

$$S_I = \sigma_0$$

$$S_{II} = (\text{sum}_{II})^{1/2}$$

$$S_{III} = (\text{sum}_{III})^{1/3}$$

For numerical examples, principal stresses are related to invariants thus:

	σ_1	σ_2	σ_3			S_I	S_{II}	S_{III}	
First example:	18	14	10	MPa	$\rightarrow$	14	5.7	0	MPa
Second example:	18	17	10		$\rightarrow$	15	6.2	5.4	

In an exactly similar manner, strains at a point can be expressed by ε_I, ε_{II}, and ε_{III} and strain rates by E_I, E_{II}, and E_{III}. In an isotropic elastic material, the bulk modulus relates ε_I to S_I and the shear modulus relates ε_{II} to S_{II} (and also ε_{III} to S_{III} if the stress–strain behavior is linear). The equations

$$S_I = 3K\varepsilon_I$$

$$S_{II} = 2G\varepsilon_{II}$$

are almost indistinguishable from eqns. (7.4) and (7.5). But viscous and diffusive behavior bring a more interesting point to light.

In viscous materials, the change-of-shape relation

$$S_{II} = 2NE_{II}$$

resembles the corresponding elastic relation, but the change-of-volume relation is not so simple. A sample of gas can change volume at constant mass by changing density, and there may be circumstances in which it does this at a rate that is linked to the current pressure. Then a relation

$$S_I = 2QE_I$$

might exist, as suggested by eqn. (7.6); here Q would be a kind of viscosity limiting the rate of change of volume. More importantly for present purposes, a sample of condensed material can instead change volume at constant density by losing some of its mass, for example by a diffusion process. This is the possibility illustrated by the brick in Figure 8.2. Here it is not the pressure itself that causes volume loss, but rather the source-plus-sink combination or the fact that the pressure is higher at some locations than at others. Formally, the effect is expressed in second derivatives and particularly in the Laplacian operator ∇^2, as in eqn. (8.7).

The point that emerges concerns symmetry and vectors. The quantities discussed above, namely stress, strain, and strain rate are all symmetrical in the manner of Figure 6.1. They are tensors of second rank and have no *direction* associated with them, in contrast to vectors. Similarly, the Laplacian has no direction associated with it, in contrast to gradients $\partial/\partial x$ and so on. One cannot believe an equation if the terms on one side have directional attributes and the terms on the other side do not; one cannot mix terms in this way. But as already noted

$$E_I = \frac{S_I}{2Q} \qquad \text{and} \qquad E_I = \frac{\nabla^2 S_I}{\cdots}$$

are both acceptable forms.

The argument that if one side of an equation has no directional attributes, the other side cannot have them either is of great power, and it applies most readily to equations among invariants. In Chapters 9 through 18, quantities that depend on direction are extensively discussed because they are less abstract and easier to visualize than the invariants. But in Chapter 19 the book's main proposals will be checked using ideas from this appendix.

(Comparison with other texts: the essence of the invariants S_I, S_{II}, and S_{III} is their independence of any directions or coordinate frame. But any arithmetic multiples have the same independence, and so do combinations such as $3S_I + 6S_{II}$. Hence many invariant quantities other than the three defined here are in use. But no system can use more than three independent invariants at a time, and any set of three is convertible into any other set. The same conclusions about possible or impossible material behavior are reached regardless of which set is used.)

9

CHEMICAL POTENTIAL UNDER NONHYDROSTATIC STRESS

The purpose of this chapter is to continue the unification that was begun in Chapter 8. There, first and second derivatives of normal stress with respect to orientation were used; we now examine the idea that the chemical potential of a component at a point can be a multivalued direction-dependent scalar like the normal-stress magnitude, and that it too can have a gradient with respect to orientation.

Associated equilibrium states

The essence of a nonhydrostatic stress is that different planes through a point are subject to different normal compressive stresses: σ_n varies with the orientation of the plane considered. Let us focus on a plane i across which the normal compressive stress is σ_i: then we put forward the assertion that an equilibrium state that can be associated with plane i is a hydrostatic state whose pressure has the same magnitude as σ_i. For illustration, see Figure 9.1.

(For the present, we take a cautious stance: each hydrostatic state in the figure is certainly an equilibrium state, and each is certainly associated with a plane, but is it *the* associated equilibrium state that *properly belongs* with that plane according to the precepts of, for example, de Groot (1951, p. 11)? For now, we make no attempt to prove that it is so: we simply use the assertion and explore its consequences. Fortunately its consequences include large amounts of classical mechanics so that it counts as a "successful assertion" on those grounds, but at least for now it lacks any underpinnings.)

An immediate consequence of the assertion illustrated in Figure 9.1 is the relation in Figure 9.2. For equilibrium states $\partial\mu/\partial p = V$, the unit volume of the material, and so for two states that differ only in the magnitude of p,

$$\mu_1 - \mu_2 = V(p_1 - p_2) \tag{9.1}$$

(neglecting any variation of V with p). Or, if successive states are identified by the magnitude of a variable α,

$$\frac{d\mu}{d\alpha} = V\frac{dp}{d\alpha} \tag{9.2}$$

So we come to a situation very like the one illustrated in Figure 4.1, the long bar: the real situation is a nonequilibrium situation (of nonhydrostatic stress),

66

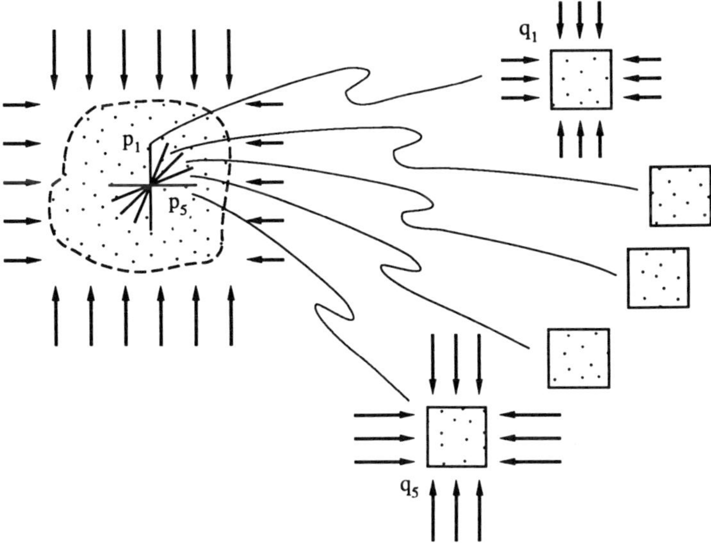

Figure 9.1 A volume-element of material in a nonhydrostatic stress state, five imagined planar elements inside the volume, and five hydrostatic conditions each of which is associated with one of the planes. The hydrostatic pressure q_1 is equal to the compressive stress p_1 on the specific plane shown, and correspondingly for the four other hydrostatic conditions indicated.

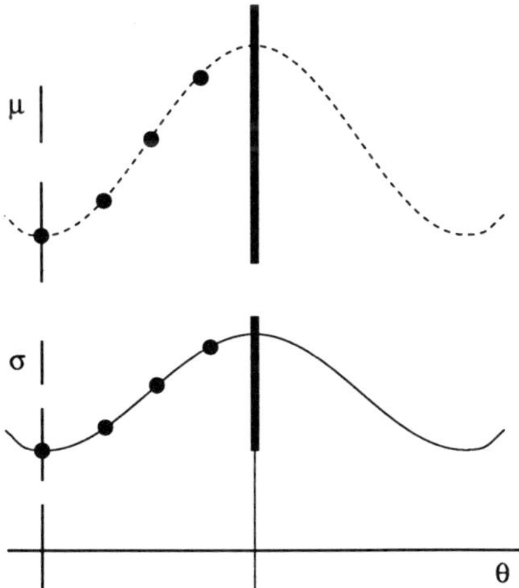

Figure 9.2 Variation of normal-stress component σ with direction θ and a set of magnitudes of chemical potential μ. Each value of μ is associated with a value of σ by an act of the imagination as in Figure 9.1.

but for each of several subdivisions we can assign a potential to an associated equilibrium state and so see a gradient in the set of associated equilibrium potentials (Figure 9.2, upper curve). Then we propose that the response in the real nonequilibrium state can be estimated from the gradient so constructed.

What will that response be? We have already explored the material's response to the variation of σ with θ (Figure 9.2, lower curve). To the extent that the procedure just used is justified, of associating a potential μ with every normal stress σ_i, the response can be seen just as well as a response to the gradients in the upper curve. Equations corresponding to (7.10) and (8.11) are

$$e_{ij} = \frac{1}{4NV}\frac{\partial \mu_{ii}}{\partial \theta_k} \tag{9.3}$$

and

$$e'_{ii} = \frac{-1}{12NV}\left(\frac{\partial^2 \mu_{ii}}{\partial \theta_j^2} + \frac{\partial^2 \mu_{ii}}{\partial \theta_k^2}\right) \tag{9.4}$$

Equation (9.4) is a second milestone and calls for two comments. First we note that it almost completes the unification sketched in Figure 8.1 where ideas from chemistry and from mechanics are on converging paths: its left-hand side, a strain rate, belongs in continuum mechanics, whereas its right-hand side is a function of the material's chemical potential. The equation at least begins to combine mechanics with chemistry and is the first equation in the book to do so. It is the business of succeeding chapters to go on from eqn. (9.4) and develop more widely applicable equations of this concept-bridging type.

The second comment concerns chemical potential itself and the deceptiveness of Figure 9.2. The two curves in that figure are related very simply by the factor V, the material's unit volume, and so are eqns. (8.11b) and (9.4). In the equations and the figure, the potential μ_{ii} and the stress component σ_{ii} seem to play very closely similar, almost interchangeable roles. But there is this important difference: the stress magnitudes in Figure 9.2 are all in some sense real and can coexist, whereas the potentials are imaginary and cannot. It is helpful to refer back to Figure 4.1: all the concentrations represented by the top profile in that figure are real and can coexist, whereas the potentials in the lowest diagram are created by five separate acts of the imagination, as the intervening diagram suggests. In the same way, an intervening diagram could be inserted in Figure 9.2: it would show five separate horizontal bars, i.e., conditions where σ does not vary with θ, the same five *separate imaginary* conditions that are shown in Figure 9.1.

The idea that, at a single point, a material actually has a range of potentials is a difficult idea and is not put forward here. The present proposal is that we can *imagine* states as in Figure 9.1, and that imagining them leads to useful real-world predictions. Chapter 5 about internal variables is relevant as well as Chapter 4: it is a reminder that one can imagine five separate possible states of a material without necessarily imagining five separate locations in space. To compare the orientation variable θ with an internal variable such as degree of polymerization may be helpful. The orientation variable is certainly not a spatial variable, but it is not quite like other internal variables either.

Sine-wave, Mohr-circle, and ellipsoid representations

The correspondence shown in Figure 9.2 can be extended to all features of Figures 6.5, 6.6, and 6.7; see Figures 9.3, 9.4, and 9.5. With Figures 6.5, 6.6,

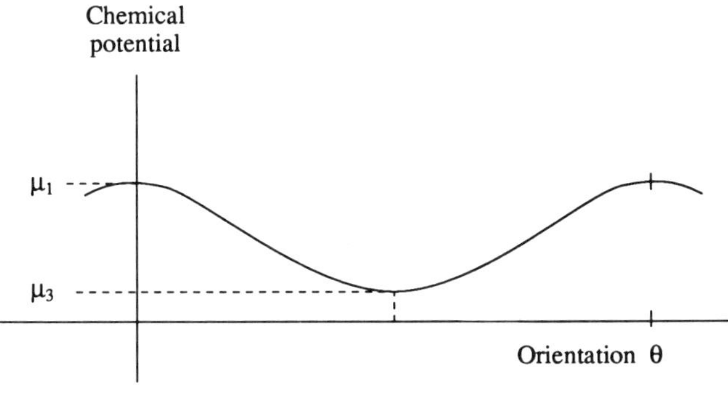

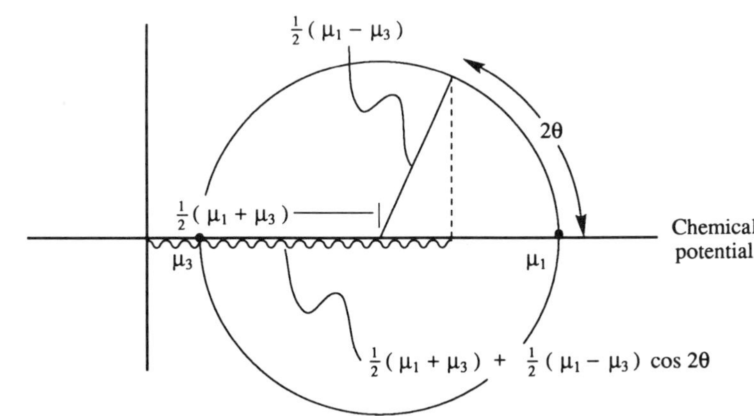

Figure 9.3 Representations of the values of the chemical potential μ established by the indirect procedures in Figures 9.1 and 9.2. The variation shown in μ is closely related to the variation of compressive stress shown in Figure 6.5, but the zero point on the μ-axis is not closely related to the zero point on the σ-axis in Figure 6.5.

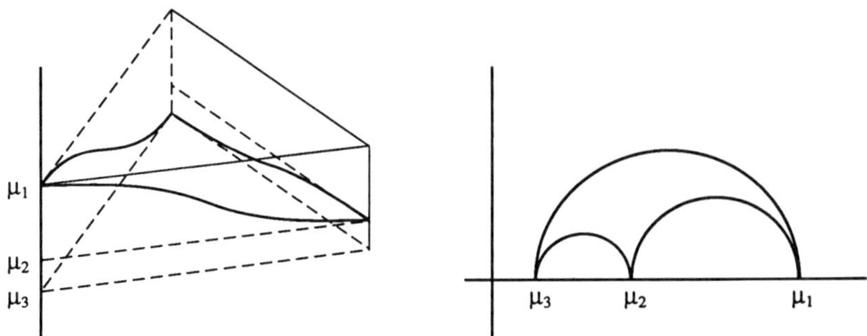

Figure 9.4 Representation of the values of μ in the manner of Figure 6.6.

and 6.7 in hand, we note that for every distance in those figures that represents a stress difference ($\sigma_1 - \sigma_2$ or similar), a corresponding diagram can be drawn containing a potential difference $\mu_1 - \mu_2 = V(\sigma_1 - \sigma_2)$. The zero-points on the μ axes are considerably displaced—and it does not really matter where they are displaced to, because it is differences or gradients of potential that drive processes; the absolute magnitude of a potential is never of any importance. So we arrive at Figures 9.3, 9.4, and 9.5, where Figure 9.5 in particular shows the equilibrium chemical potential associated with any plane in the material, as identified by its normal $\mathbf{n}$; stresses and strains can be positive or negative, but μ is always positive and hence a diagram like Figure 9.5 can always be drawn: it is an ellipsoid of equilibrium chemical potentials defined by the three principal equilibrium chemical

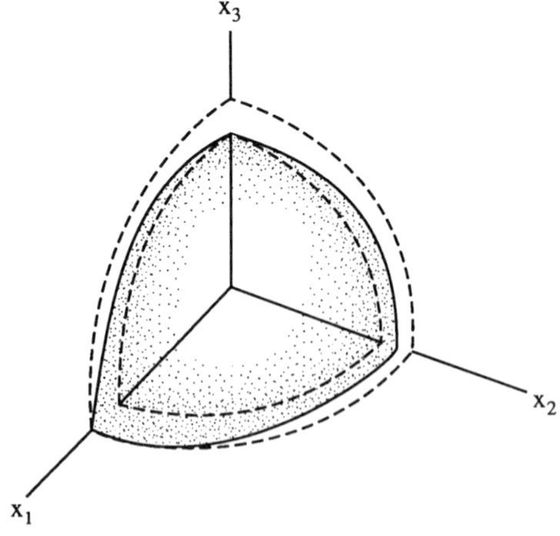

Figure 9.5 Representation of the values of μ in the manner of Figure 6.7.

potentials μ_1, μ_2, and μ_3, and by the relation

$$\mu_n = n_1^2 \mu_1 + n_2^2 \mu_2 + n_3^2 \mu_3 \tag{9.5}$$

Numerical examples

Let us consider a stress state with principal stresses 18, 14, and 10 MPa. In the 1,3 plane $\sigma_n = 14 + 4 \cos 2\theta$; then $\partial^2 \sigma_n / \partial \theta^2 = -16 \cos 2\theta$. Representative magnitudes are:

$$\theta = \begin{array}{cc} 0 & \partial^2 \sigma_n / \partial \theta^2 = -16 \text{ MPa/(radian)}^2 \\ 30 & -8 \\ 45 & 0 \\ 60 & 8 \\ 90 & 16 \end{array}$$

If the material's volume is $0.0004 \text{ m}^3/\text{kg}$, corresponding magnitudes for $\partial^2 \mu_n / \partial \theta^2$ are -6400 J/kg-rad^2, etc. If the material's molecular weight is 60, corresponding molar quantities are molar volume $0.024 \text{ m}^3/\text{kg-mol}$ and second derivative $-384 \text{ kJ/kg-mol-rad}^2$.

Turning to the 1,2 plane, here $\sigma_n = 16 + 2 \cos 2\theta'$; a corresponding series of second derivatives could be calculated, all half as large as above, starting with -3200 J/kg-rad^2. If a diagram like Figure 9.4a were drawn, all these second derivatives would appear, more or less, as curvatures; they would not appear exactly because the length-scales in the diagram are related to θ in a weakly nonlinear way. In a diagram like Figure 9.5, each curvature seen is a composite formed by combining a curvature term like $\partial^2 \mu / \partial \theta^2$ with the curvature of the reference sphere, of radius $\mu_0 = (\mu_1 + \mu_2 + \mu_3)/3$. But at the x_1-extremity, for example, one can see that if the stippled surface was properly constructed in three dimensions, its curvature would deviate from that of the reference sphere by twice as much in the 1,3 plane as in the 1,2 plane.

To go on and use eqn. (9.4) we need the material's viscosity: let this be 2×10^{10} Pa-sec, like some waxes or red-hot glass. Then the linear strain rate along direction 1 is $(6400 + 3200)/12(2 \times 10^{10})(0.0004)$ or 10^{-4} per sec; the material would be shorter by about one-third after an hour. If, instead of eqn. (9.4), we use the regular continuum mechanics equation (7.7), $e' = \sigma'/2N$, we need to note that the mean stress $(\sigma_1 + \sigma_2 + \sigma_3)/3$ is 14 MPa, so that σ_1' is 4 MPa. Then e' comes to 10^{-4} per sec as before.

Summary

If a material is in a state of nonhydrostatic stress, and at some point in the material we identify a direction by a unit normal vector $\mathbf{n}$, then with the vector $\mathbf{n}$ and the plane it is normal to, there is associated an equilibrium chemical potential, designated μ_n and following eqn. (9.5).

Equilibrium potentials are closely related to normal stresses: $\partial\mu_n/\partial\sigma_n = V$, the material's unit volume.

The material's response to nonhydrostatic stress, especially its principal strain rates e_1, e_2, and e_3 and associated strain rates in nonprincipal directions, can alternatively be seen as a response to the set of equilibrium chemical potentials. Relations of the type

$$e_n \quad \text{or} \quad \frac{d\varepsilon_n}{dt} = -k\frac{d^2\mu_n}{d\theta^2}$$

exist. On the right, orientation is used as an internal variable, $d\theta$ being a small change in the direction of the vector $\mathbf{n}$.

As so far introduced, the description of material behavior in terms of μ simply reproduces the description in terms of stresses and adds nothing new; but when a problem is encountered where material is driven simultaneously by a stress field and, say, a concentration gradient, the ability to express both influences in terms of a single variable μ will bring advantages over any approach that retains the stresses in explicit form.

II
SIMULTANEOUS DEFORMATION AND DIFFUSION

10

INTRODUCTION

In Chapters 2 through 9, ideas were introduced as follows.

(A1) When considering movement of material, a useful idealized relationship is of the form

$$\text{Flux along direction } x = -k\,\frac{\partial(\text{potential})}{\partial x} \qquad (10.1)$$

(A2) Movement of material occurs in nonequilibrium situations, whereas potentials are defined with fewest complications in equilibrium situations.

(A3) Ideas (1) and (2) can be fitted together by finding, for any actual nonequilibrium situation, a set of imaginary associated equilibrium states. If well chosen, the imaginary states provide potentials that, by entering the right-hand side of eqn. (10.1), serve as a guide to the flux in the real state.

(A4) Where fluxes follow eqn. (10.1), conservation statements take the form

$$\frac{\partial Q}{\partial t} = k\,\frac{\partial^2(\text{potential})}{\partial x^2} \qquad (10.2)$$

where Q is the quantity of material in a unit volume, or its local density at the point of interest.

As an example, the preceding ideas can be applied where a minor component of a mixture diffuses through its host under the influence of a potential gradient that exists because of nonuniform composition.

A second group of ideas arises from considering the deformation of viscous materials.

(B1) Viscosity N is defined by

$$\text{Shear stress} = 2N \times (\text{shear strain rate}) \qquad (10.3)$$

(B2) If we conserve material and momentum, it follows from the definition (10.3) that

$$\frac{\partial(\text{linear strain})}{\partial t} = \frac{-1}{12N}\left(\frac{\partial^2\sigma_{ii}}{\partial\theta_j^2} + \frac{\partial^2\sigma_{ii}}{\partial\theta_k^2}\right) \qquad \begin{array}{c}(10.4)\\{[=8.11b]}\end{array}$$

Here σ_{ii} is a normal-stress component and θ is a variable specifying the orientation of direction i.

(B3) If, for any plane on which the compressive stress is σ_{ii}, we imagine an associated hydrostatic state p, where $p = \sigma_{ii}$, then we can establish a chemical potential μ_i and rewrite eqn. (10.4) in the form

$$\frac{\partial(\text{linear strain})}{\partial t} = \frac{-1}{12NV}\left(\frac{\partial^2 \mu_i}{\partial \theta_j^2} + \frac{\partial^2 \mu_i}{\partial \theta_k^2}\right) \qquad (10.5)$$

Here V is the material's unit volume.

There are obvious parallels between eqn. (10.5) and eqn. (10.2); in fact eqn. (10.5) suggests that continuum mechanics is just one more set of applications of the idea that materials tend to move down gradients of chemical potential. But there is also a conspicuous difference: ideas in group A embody the idea that, at any point in space, a material component just has a chemical potential—a single value; by contrast, group B embodies the idea that, under nonhydrostatic stress, it is a plane or a direction i that has a chemical potential associated with it, and that the potential associated with one plane can be different from the potential associated with another plane at the same point in space. The objective of this book is to treat situations where deformation and interdiffusion are occurring simultaneously; that is to say, we want to combine eqns. (10.2) and (10.5), and hence the question, "Is chemical potential single-valued or multi-valued?" must be faced.

The purpose of the ensuing Part II is to work on the question just set out. It will be proposed that, in general, a component's chemical potential at a point is multivalued (being single-valued only in the special case of hydrostatic stress). In other words, eqn. (10.5) correctly shows a material's linear strain being related to the second derivative of its potential with respect to a variable θ that specifies orientation; it is eqn. (10.2), relating the quantity of material in a cell to the second derivative of its potential with respect to position in space, that needs to be examined and refined. When refined, its resemblance to eqn. (10.5) will be even closer.

Novel features of the treatment presented in this book are:

1. Use of θ, the variable for orientation, and of second derivatives with respect to θ. These practices lead to the shortening-rate equations (10.4) and (10.5), which show second derivatives of potential with respect to orientation. These are alternatives to the shortening-rate equations used in classical continuum mechanics. The latter are of course equally *correct*, but for present purposes less suitable in form than eqn. (10.5).

2. Emphasis on the resemblance between the terms $\partial^2(\text{potential})/\partial\theta^2$ and the terms $\partial^2(\text{potential})/\partial x^2$. It will be argued that this is not accidental; on the contrary, the resemblance must exist—it would be unnatural if it did not. It is by paying attention to this resemblance that we gain, in Chapters 11 and 12, a highly unified description of a continuum's behavior. Self-diffusion from sites of high compression (sources) to sites of low compression (sinks) combines with a material's tendency to change

shape. Just one field of potentials exists, whose gradients with respect to position give one set of behaviors and with respect to orientation give the other set.

Following the generalities in Part II, some specific physical situations are treated in Part III, and readers whose taste is toward the concrete may wish to skip ahead. If one wonders where Part II leads and what its consequences are, it is Part III that one needs. The ideas covered in Part II are fundamental, but they are necessarily abstract and possibly cloudy. To explore Part III while taking Part II temporarily on trust is a reasonable alternative to progressing straight through.

Historical review: descriptions of diffusive mass transfer accompanying deformation

This section is a selective, not an exhaustive, review.
Situations that have been considered are:

1. A permeable elastic host and low-viscosity fluid (Terzaghi 1925, Biot 1941).
2. A mass of grains, each wetted by a boundary layer of fluid; the material of any grain can dissolve in the fluid and diffuse away to be precipitated at another location, the fluid itself remaining stationary throughout (Fletcher 1982).
3. An amorphous stiff fluid mixture with atoms of two species A and B (Stephenson 1988).

1. *Permeable elastic host.* The first class of situations includes as an example the situation in Figure 8.3a. Suppose that at the moment of loading, the elastic host is fully relaxed, i.e., free of stress: then if we exclude inertial effects, at the moment of loading, the load is opposed wholly by pressure in the fluid. (*After* fluid has traveled and the elastic host has deformed, stresses in the host build up and support part of the load, but at the first moment, this effect is zero.) Then, focusing on the fluid, it is customary to imagine several things at once: (i) the movement of fluid is sufficiently slow for inertial terms to be neglected, as already mentioned; (ii) the reason for the movement being slow and the force–flux relation being linear, as in eqn. (8.3), is that the fluid moves along narrow tortuous channels where its viscosity dominates its flow behavior; (iii) despite the flow on the scale of the tiny channels being controlled by shear stresses (i.e., nonhydrostatic conditions), the average state of the fluid in one part of the channel can be described by a hydrostatic pressure P, whereas farther along the channel it can be described by a different hydrostatic pressure Q. Success in practical engineering situations comes from assuming that a field of single-valued pressures, P, Q, etc., controls the flow, even though the force–flux relation used depends on the compressive stress at a point not being single-valued.

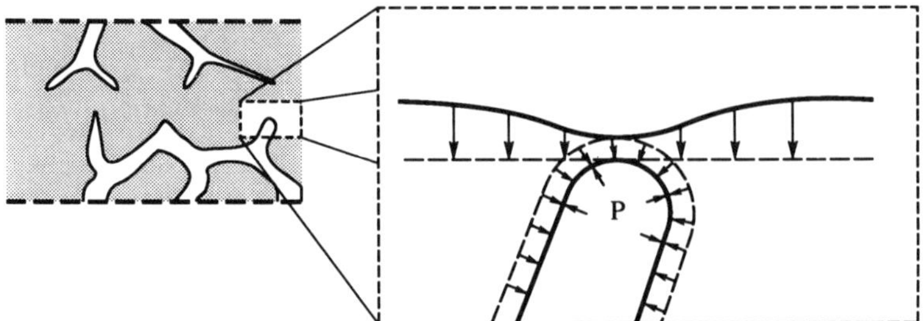

Figure 10.1 An elastic material with interconnected pores. If the pore fluid is at pressure P, the stress in the material must match the fluid pressure P at every interface, but stress components in the interior of the material can differ from P.

At the other extreme, after an infinite time, the load is supported wholly by the elastic host, the fluid stops flowing and the fluid pressure is not only hydrostatic at each local point but spatially uniform as well. This means that at all fluid–solid interfaces, the stress in the solid has a component normal to the interface of some magnitude P (see Figure 10.1); it is only in the solid's interior, away from any flat-lying fluid interface, that vertical compressive stresses with magnitude greater than P can exist and support the load.

At times intermediate between first loading and stagnation, the stress state will be intermediate also. It is described by means of a local average stress state in the elastic host that is nonhydrostatic, and a local average stress state in the fluid that is hydrostatic; the fluid pressure varies from point to point in the same sinusoidal manner as at the first moment but with diminished amplitude.

2. *Low-porosity mass of soluble grains.* The situation discussed by Fletcher differs from the above in important ways. Again we have two phases, a mechanically robust host-material and a small volume-fraction of an interstitial fluid; but here the host-material is viscous rather than elastic (with extremely high viscosity), and the interstitial fluid has no ability to move from place to place—it is more like a stationary surface film that functions only as a solvent and a channelway for the diffusion of solute. Under a load such as is shown in Figure 8.3a, effects are: (i) the viscous host deforms as a continuum (see Figure 10.2), and also (ii) host material

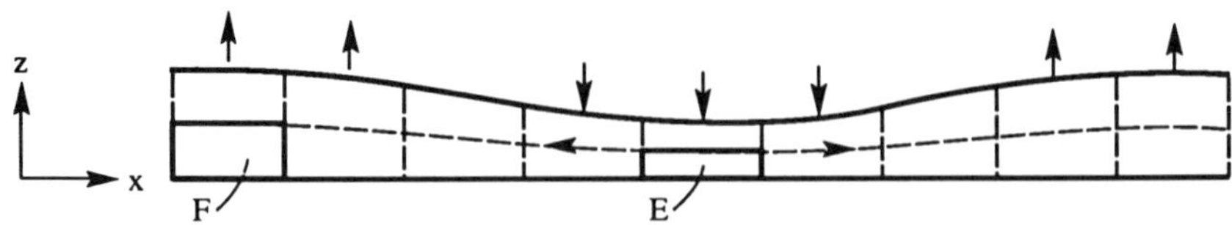

Figure 10.2 Deformation of a layer that was formerly of uniform thickness, under a nonuniform vertical load.

dissolves at high-pressure sites, diffuses along the interstitial channelways, and reprecipitates at low-pressure sites. The dimensions of a volume element such as E thus change for two reasons: there are volume-conserving strain rates e_{xx} and e_{zz} where $e_{zz} = -e_{xx}$, magnitudes being governed by the viscosity and local stress difference $\sigma_{zz} - \sigma_{xx}$; in addition, there are strain rates produced by solution activity that add up to an overall volume-loss from elements such as E and gain by elements such as F. The *solute* thus moves from place to place as the *fluid* did in situation (1) above.

One may ask what drives the flux of solute. In Fletcher's theory the direct cause is a nonuniform chemical potential linked to a non-uniform concentration of solute; he argues that the concentration will be higher in cell E than in cell F. The reason for the concentration being higher in cell E is the stress state; it is assumed that the material's solubility is higher where the pressure is higher. Now at cell E, the vertical compression is certainly greater than the horizontal compression (hence the flattening strain rates e_{xx} and e_{zz}) and it is compression that drives the dissolution process—yet Fletcher takes the shortening strain rates produced by dissolution to be equal along x and z: the volume-loss is taken as an isotropic shrinkage despite the driving stress state being noticeable anisotropic.

Fletcher's example is of great importance. It is the clearest, simplest treatment of a situation where only one material moves, but moves by two separate mechanisms—a volume-conserving viscous change of shape plus a diffusive mass transfer from higher-compression sites to lower-compression sites. The two mechanisms resemble those operating in situation (1), the elastic permeable host plus mobile fluid; Fletcher clearly brings to light the fact that the two movement mechanisms remain distinct even when the same material travels by them both.

3. *Amorphous stiff fluid mixture.* Stephenson considers a stress state again resembling the one shown in Figure 8.3a; like Fletcher, he recognizes that material will move by two mechanisms, a volume-conserving change of shape and a diffusive mass transfer; again like Fletcher, he takes the diffusive mass transfer to be driven by gradients in chemical potential. Differences are, first, that the simplest of Stephenson's materials is an amorphous mass of atoms (like a heat-softened glass) and, second, that Stephenson does not involve any explicit step like the dissolution process in Fletcher's discussion: Stephenson simply assumes that atoms have two behaviors—on one hand they cluster together to form a mass that mechanically approximates a continuum, and simultaneously on the other hand they act as individual diffusers that tend to travel down gradients in chemical potential. In Stephenson's treatment, as in Fletcher's, diffusive mass transfer takes material from higher-pressure sites to lower-pressure sites, but with Stephenson the gradient in chemical potential goes *directly* with the gradient in compressive

stress, following

$$\frac{\partial \mu}{\partial x} = V \frac{\partial p}{\partial x} \tag{10.6}$$

as in Figure 9.2. (With Fletcher, the stress gradient in the mass of grains produces a concentration gradient, and the concentration gradient gives the chemical potential gradient in the interstitial fluid film or diffusion channelway.)

But again the assumption is made that any change of volume produced by diffusive mass transfer is isotropic. Stephenson, like Fletcher, assumes that even though it is compression that causes volume loss, the shortening of a cell's linear dimension by volume-loss will be the same along the direction of least compression as it is along the direction of greatest compression. This is certainly a *simple* assumption that permits rapid progress to the next stage, of the exploring the theory's consequences; but it seems to have little other than simplicity in its favor and prompts the question, "What else might we assume instead?"

The early part of this chapter emphasized the question, "Can the chemical potential of a component at a point in space be multivalued and direction-dependent, or must it be single-valued?" As long as a component's chemical potential is taken as single-valued, it is perhaps hard to see why any volume-loss driven by the potential should be other than isotropic. But if we admit the possibility of a component having one potential associated with the plane of maximum compression and a *different* potential associated with the plane of minimum compression, etc., clearly this opens the way to finding a reason for the shortening by volume-loss to be different from one direction to another. Thus the question about Stephenson's simple assumption and the main question, "single-valued or multivalued?" are linked; the short historical review just given brings us back to the main question this chapter introduces.

As foreshadowed in Chapter 1, the suggestions of this book are that, even at a single point, a component's chemical potential is multivalued; in a nonhydrostatic stress state, if volume-loss occurs at a point, it does not occur by equal shortenings in all directions; volume-loss is controlled by second derivatives of the form $\partial^2(\text{potential})/\partial x^2$ but separate relations exist for the three principal directions:

$$\frac{\partial \varepsilon_{11}}{\partial t} = \frac{k}{3} \nabla^2 \mu_1 \tag{10.7a}$$

$$\frac{\partial \varepsilon_{22}}{\partial t} = \frac{k}{3} \nabla^2 \mu_2 \tag{10.7b}$$

$$\frac{\partial \varepsilon_{33}}{\partial t} = \frac{k}{3} \nabla^2 \mu_3 \tag{10.7c}$$

Since

$$\partial(\varepsilon_{11} + \varepsilon_{22} + \varepsilon_{33})/\partial t$$

equals the volume strain rate $\partial(\text{volume})/(\text{volume}).\partial t$, these add up to give the overall effect assumed in the isotropic approach:

$$\frac{\partial v}{v\,\partial t} \quad \text{or} \quad \frac{\partial Q}{\partial t} = k\nabla^2\mu_0 \qquad \text{where } \mu_0 = \tfrac{1}{3}(\mu_1 + \mu_2 + \mu_3) \qquad (10.8)$$

per unit volume

Thus the set of equations (10.7) is a *possible* set compatible with (10.8), but they are not the only such set; other groups of three could be devised that add up to eqn. (10.8) just as well. Chapter 11 presents an argument that the equations (10.7) are the correct set, but again, readers who want to see consequences rather than underpinnings are invited to skip ahead.

11

DEFORMATION AND DIFFUSION COMPARED

The suggestion in view is that when volume is lost by diffusive mass transfer, the consequent shortening rate along some direction $\mathbf{n}$ is controlled by $\nabla^2\sigma_{nn}$ regardless of the spatial variations in other stress components. The nature of the argument advanced is comparable with the one on which the theory of relativity is based: "At two separate points in a universe, it is not reasonable to suppose that the fundamental laws of behavior will be different at one point from the other." If it is *only* in respect to some reference frame set up by an observer that point P differs from point Q, one should not expect behavior at P to differ from behavior at Q. It is convenient to use anthropomorphic phrasing: "If there is nothing intrinsic about point P to tell the material there to behave differently, the material at P will behave in the same way as the material at Q."

The theme of this chapter is that the material process for diffusive mass transfer is almost indistinguishable from the process for volume-conserving viscous change of shape at a point. In fact it will be argued that the two processes are *so similar* that it is not reasonable to suppose that behavior will be governed by different laws in the two modes: only an *observer* can distinguish one process from the other. Again anthropomorphically, "The moving material itself has no means of knowing which process it is involved in. Hence, if it is direction-dependent quantities such as σ_{nn} that control behavior in change of shape at a point, it must also be direction-dependent quantities such as σ_{nn} that control diffusive mass transfer."

In presenting the argument, it is convenient to imagine an atomic material for purposes of *example*, and for the sake of concreteness; but it is emphasized at the outset that the atoms are of minimal significance—the objective is a theory for a continuum. We wish to treat a continuum in which diffusion occurs, and even a continuum with only one component in which *self-diffusion* occurs, and most people find that this requires imagining division of the continuum into particles on some scale: but we need this division only in the most abstract sense, just enough to permit the idea that the continuum is self-diffusive. Any other attributes of atoms that are used in the following examples are purely for concreteness, to help the imagination, and are not an intrinsic part of the theory.

Preliminary sketches

Subject to the proviso just given, we turn to Figure 11.1. The intention is to suggest an extensive three-dimensional lattice of atoms and to focus on a small planar group shown by twenty-five open circles in diagram (a); these are shown again edge-on in diagram (b). The stress normal to the planar group is σ_1; a smaller stress σ_2 acts on plane P, and also at a short distance to one side, an equal smaller stress σ_2 acts in the same direction as σ_1 on an area Q; the compressive stress is taken to diminish with position across the space shown in some smooth way from σ_1 to σ_2.

The point to be made using Figure 11.1 is that if the 25 atoms can migrate either to site P or to site Q, the work done by the stress σ_1 and the work done against the stress σ_2 are much the same in either case. One is free to imagine any kind of transport process, such as movement of single atoms by formation of Schottky defects; the *mechanism* is of no interest, only the

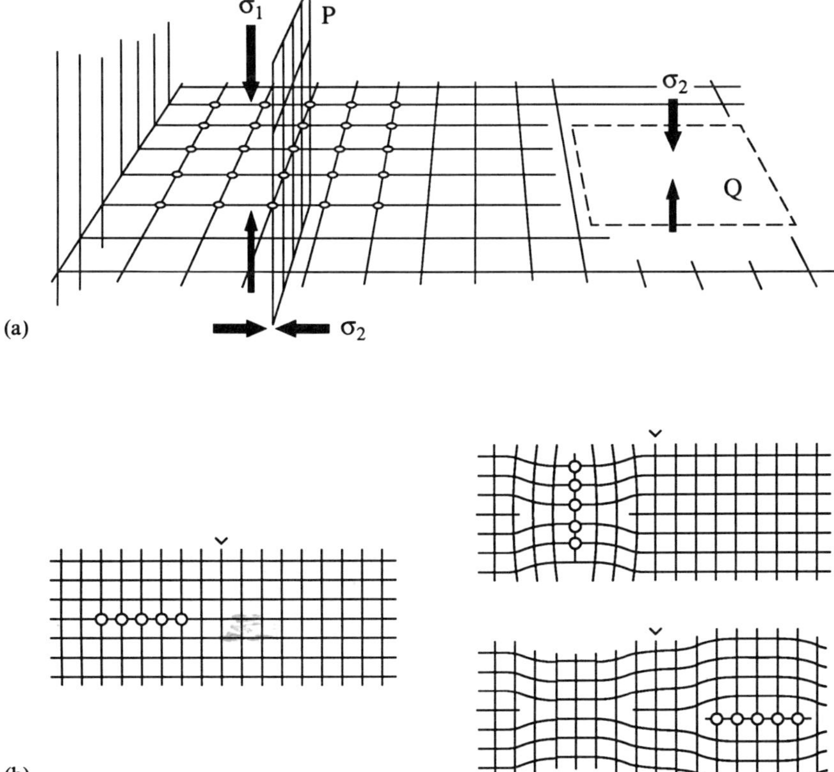

Figure 11.1 Part of a sample of material composed of atoms on a cubic lattice. Every lattice point carries an atom but twenty-five atoms are emphasized for purposes of discussion in the text.

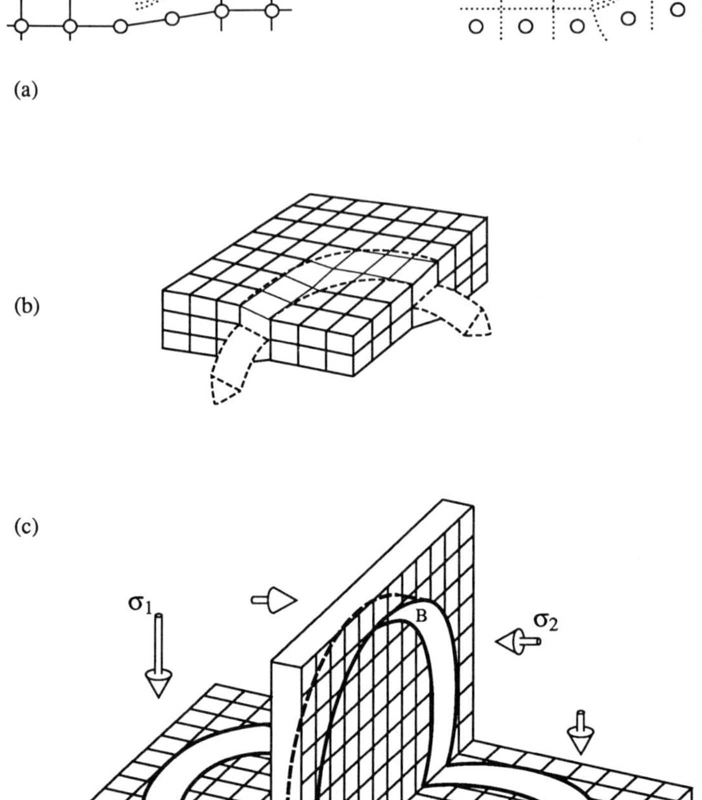

Figure 11.2 (a) An edge dislocation shown in cross-section. The dislocation core is idealized as a triangle. (b) A curved dislocation. (c) Three curved dislocation loops intersecting.

fact that for the 25 atoms, the initial state, the final state, and the work done are much the same in either of the migrations.

A slightly different picture with many parallel features is shown in Figure 11.2. The figure shows three idealized dislocation loops, each loop being the edge of an incomplete lattice plane. Again we have a portion of the system subjected to compression σ_1 and two portions subjected to compression σ_2. Again it is possible for material to escape the compression σ_1

and migrate to either of the less strongly compressed sites. It is easy to suppose that the three loops are equally easy diffusion pathways. Whatever the relation between the stress-drop $\sigma_1 - \sigma_2$, the length of the diffusion path from source to sink, and the rate of migration of material, we begin to see grounds for proposing that the same relation must govern migrations A → B and A → C *because atoms have no sense of geography.* An *observer* can see that loops A and C are coplanar whereas the journey A → B involves a right-angle turn, but an atom has no sense of this distinction: *it is not reasonable to suppose that different force–flux relations govern migrations A → B and A → C,* if the difference between them depends on having a sense of where "north" is, or "up," or some sense that atoms do not possess.

The discussions of Figures 11.1 and 11.2 are merely preliminary. As emphasized, the atomic details are not essential parts of the theory. We wish to proceed to a treatment that does not depend on any atomic details and at the same time is more mathematically complete, and this is attempted in connection with Figure 11.5 below. As one more preliminary discussion, we consider the "stadium analogy" as follows.

Imagine a sports stadium enclosing a space that is an elongated loop, Figure 11.3. High tiers of seats are provided for spectators, and access to these is through windowless passages through the underparts of the structure that support the seats. Two such passages are shown at A and B, and it is a peculiarity of the stadium shown that the length of passage A equals that of passage B. Some sports event comes to an end, large numbers of people

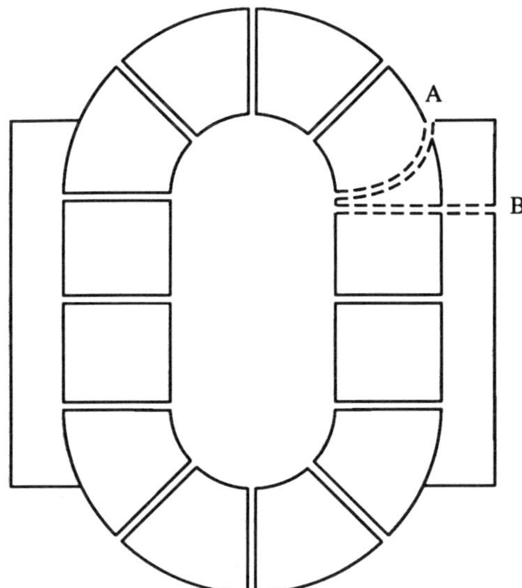

Figure 11.3 A material's migration path can be curved or straight. The diagram shows an analogy with the passages that give access to a sports stadium.

begin to shuffle and edge their way down passages A and B, and the crowd includes a child whose height is less than that of most people in the crowd. For some minutes the child shuffles and edges her way along with the rest, and finally emerges either at A or at B. While traveling, the child knows in which direction to move because of being bumped into more from behind than from in front, but she sees around her only larger people's arms, coats, etc.; while traveling she has no means of knowing whether she is in passage A or passage B. When we turn to the abstract notional particles in an ideal self-diffusing continuum, we shall imagine them traveling on paths such as A (during homogeneous change of shape) or on paths such as B (during diffusive mass transfer through space) and having, again, no ability to sense any difference between the two processes.

Continuum behavior using wafers

Our purpose is to describe self-diffusion and change of shape in a continuum. As already emphasized, we wish to avoid assuming that the material is a mass of atoms (and even more certainly, it is not a mass of shuffling children), so what mechanism shall we postulate that permits the two processes? The following description of migrating *wafers* is devised specifically to maximize the similarity between the two processes while fitting the continuum idealization. Assumptions are made as needed and review of the assumptions is deferred until later; the simplest approach is to give a bald description first, with judgment suspended.

Linear strain in one direction. The continuum has the ability to shorten along some direction z at constant density, and Figure 11.4 offers a model. The ideas are most simply presented by assuming that the material shortens simultaneously along direction y at the same rate, i.e., the shortening is cylindrically symmetrical about direction x. We imagine that during a small

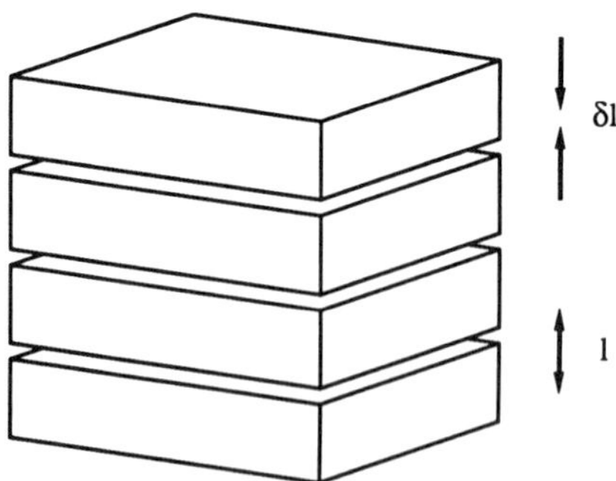

Figure 11.4 Deformation of a continuum. Lengths l and δl are identified as small finite quantities; then the ratio is kept constant as the quantities tend to zero.

finite time δt, a set of finite wafers normal to z, with thickness δl and separation l is removed to a new location. (In fact we shall assume they are removed from sites normal to z and reinserted in sites normal to x). Then we approximate continuum behavior by first letting δt and δl remain in proportion while approaching zero, and second letting δl and l remain in proportion while approaching zero. (In other words, l and δt approach zero as $1/n$ and δl approaches zero as $1/n^2$ as n tends toward infinity.) The effect is a velocity gradient or strain rate along z of magnitude $\delta l/(l\,\delta t)$.

Change of shape at constant density. One possible means by which the wafers are able to leave sites normal to z and enter sites normal to x is indicated in Figure 11.5b. Any wafer can travel by sliding sideways in its own plane, and its successive positions trace a cylindrical surface or slot of thickness δl. When first conceived, these slots start from sites with small finite separations l and cross each other without intefering; then, as above, we approach continuum behavior by letting l approach zero. But the length of the arc or quadrant does not approach zero; it remains at a finite length that is an assigned characteristic of the continuum envisaged; we designate this length by the symbol L_0.

Diffusive mass transfer. We now turn to a situation different from the one just described, but comparable with it, as in Figure 11.6. A portion of a continuum is subjected to a stress field that does not vary along y or z, but varies harmonically along x, thus

$$\sigma_{yy} = \sigma_{zz} = \tfrac{1}{2}[\sigma_1 + \sigma_3 + (\sigma_1 - \sigma_3)\cos 2\pi x/\lambda] \qquad (11.1a)$$

$$\sigma_{xx} = \sigma_3 \qquad (11.1b)$$

The situation again has cylindrical symmetry, i.e., at any position along the x-direction all material lines normal to x behave similarly. Thus, for example, parallel to direction z, we expect shortening at site (i) and elongation at site (ii), and as before the change can be conceived as occurring by wafers that lie normal to z sliding sideways in their own plane. The main difference is that the slot followed by a wafer is now planar, whereas in the previous discussion it was cylindrical. The resemblance to the stadium analogy is intentional; to maximize the resemblance, let $\lambda = L_0$ and let the principal stresses in the earlier homogeneous constrictive field be σ_1 and σ_3. Does any difference remain that could make the behavior of a wafer in Figure 11.5 different from the behavior of a wafer in Figure 11.6?

Comparison of the two situations. The sequence of compressive stress magnitudes normal to the slot that affect a wafer as it slides sideways is exactly the same in the two situations—the specifications were devised to make this so. Then *if* the process of assigning an associated equilibrium state to a plane is valid, we have to assign the same sequence of associated equilibrium states to successive points, regardless of which path we are following. Then further, *if* the gradient down a sequence of associated

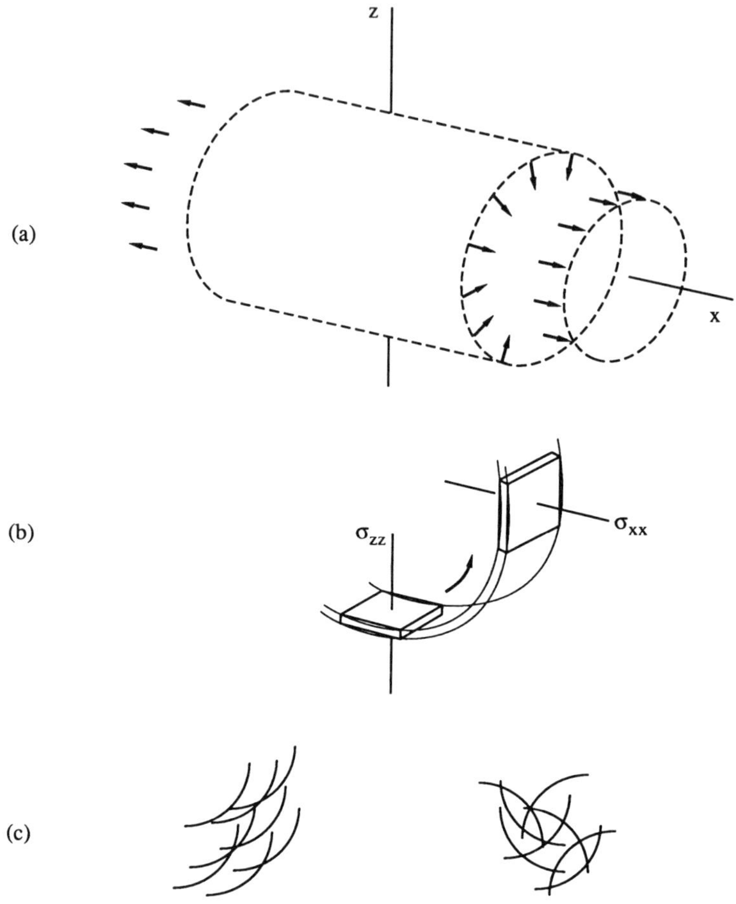

Figure 11.5 Homogeneous constriction. (a) A homogeneous deformation, with uniform radial shortening in yz planes and elongation parallel to x. (b) An imaginary wafer and a migration that would contribute to the overall deformation in (a). For continuum behavior, such wafers would need to be infinitely numerous and infinitely small, but the length of each migration path would remain finite. (c) Diagrams to suggest the concept of a dense swarm of such paths.

The diagrams illustrate specifically shortening along z. For radially symmetrical constriction, wafers normal to y would also need to migrate, or one can imagine that wafers normal to any radius all behave similarly.

equilibrium states is an effective predictor of real behavior, we would have to predict the same behavior in the slot in Figure 11.6 as in the slot in Figure 11.5. (Here we continue to envisage an ideal continuum: "real" is used just to distinguish the real nonhydrostatic state we assume the continuum to be in from the imaginary hydrostatic states needed for establishing equilibrium values.) There certainly is a gradient in mean-stress magnitude from left to right across Figure 11.6, but the remarks just made

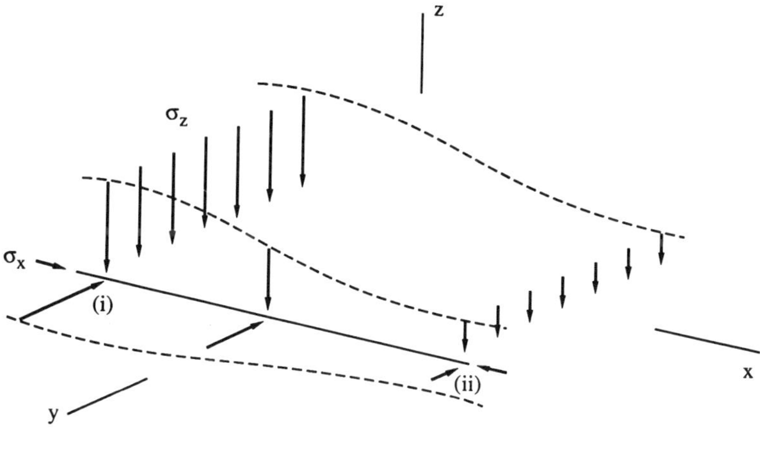

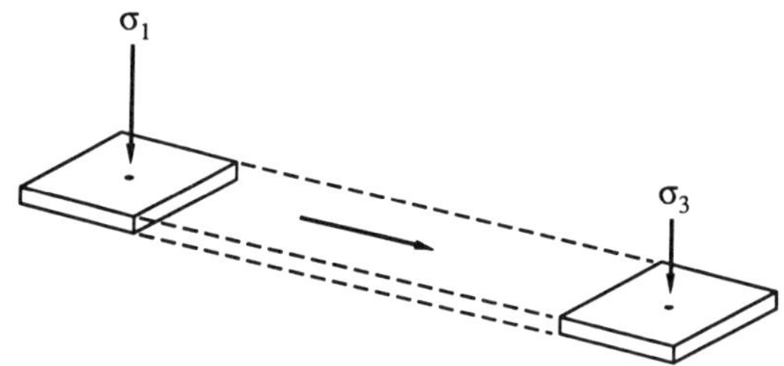

Figure 11.6 Inhomogeneous constriction. Processes at site (i) are as shown in Figure 11.5 but additional effects are present. In particular, at site (i) a second mechanism for shortening along z operates: wafers normal to z can migrate in their own plane along direction x, producing elongation parallel to z at site (ii).

are intended to show that it is not reasonable to propose that the material movement is driven by the mean-stress gradient. In Figure 11.5 it is clearly the gradient in the slot's normal-stress component that drives motion (the gradient in mean stress along the slot being zero), and hence it is reasonable to propose that the normal-stress gradient is the control in Figure 11.6 as well.

To pursue the train of thought to completion: a normal-stress component belongs with a single plane, and if we consider a set of planes through a point, there is a multivalued set of normal-stress components associated with them. A single-valued mean stress and a single-valued chemical potential go together; but if we want to relate diffusive mass transfer to chemical potential, and if it is the multivalued normal-stress components that control the transfer, then we need a multivalued set of chemical potentials. In face of the question "Is chemical potential single-valued or multivalued?" we *need* to answer that it is multivalued. The alternative is to assign the diffusing material some ability to respond differently in the two slots.

The preceding paragraph completes a sketch of the line of reasoning followed, which supports the three separate equations in the set (10.7). In the next section, weaknesses in the argument are reviewed. Appendix 11A following this chapter treats a situation similar to the one shown in Figure 11.6 that is in some ways an even closer match with Figure 11.5; and Appendix 11B offers a second, short and wholly independent argument in favor of multivalued potentials. As stated before, a reader more interested in the *use of* eqs. (10.7) than in further discussion of their misty foundations should skip ahead to Part III.

Questions about validity

Several questions can be raised about the validity of the argument given above and these are treated in succession:

1. Differences in the two sequences of stress states
2. Migration of wafers versus particles
3. Lack of realism in the continuum imagined
4. Significance of the length L_0

Differences between the stress fields. The essential point about the cylindrical slot, Figure 11.5, and the planar slot, Figure 11.6, is that the sequence of normal-stress components along the slot is the same in each; but differences are certainly present, as is readily seen on considering other stress components. In particular, consider a series of planar elements across each slot, Figure 11.7: in diagram (b), the normal-stress components on these elements are all equal, whereas in diagram (a) they are not.

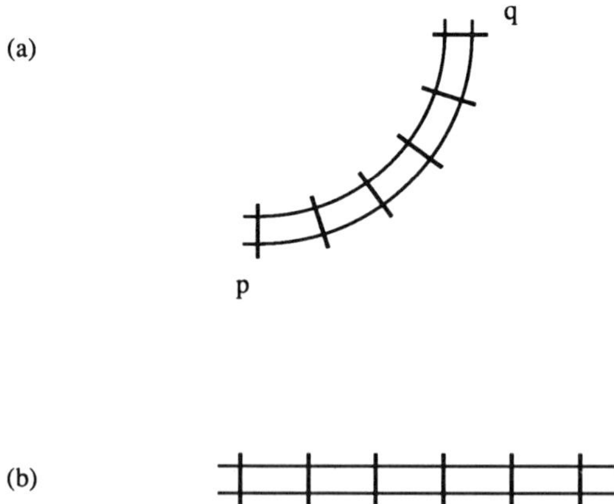

Figure 11.7 Details from Figures 11.5 and 11.6. Normal stresses on the planar elements are not similar in the two situations.

In diagram (a), in fact, the normal-stress components on the elements shown increase from p to q by just as much as the transverse components decrease; a spherical particle traveling from p to q would travel down one gradient, but would travel up another gradient equally steep. The argument offered above that a traveling *wafer* would be subject to the same physics in both slots thus hinges crucially on its being a wafer and not a particle. If the practice of envisaging wafers is valid, the *similarity* between Figures 11.5 and 11.6 is the essential point and the difference that Figure 11.7 shows can properly be ignored. Thus the stress components that Figure 11.7 brings to our notice do not directly invalidate the argument offered, but they make using wafers critical; if we did not focus on wafers, the argument would fall apart. Thus we need to move on to the next topic.

Wafers and particles. The crux of the matter is shown in Figure 11.4. As has been emphasized, the theory we are trying to build is a *continuum* theory: the velocity field is continuous and all material surfaces are continuous. It is admitted that comparable instances in the real world involve atoms, and a person's mental image of a diffusion process is apt to involve particles, but the continuum theory does *not* involve particles—yet the continuum theory *does* involve surfaces. A velocity gradient $\partial u/\partial x$ seems to derive from just two particles whose velocities differ by δu and whose positions differ by δx, but this is misleading: a velocity gradient in a continuum implies *continuity* as well as position and velocity. Thus in Figure 11.4, we do not find an isolated particle (like a gas molecule) whose distance l from another particle changes by δl; we find a particle that is part of a continuous surface whose distance l from another surface changes by δl. The tendency for part of one's mind to be concerned all the time with real atomic materials confuses the issue, and makes difficulties for the other part of one's mind that is trying to envisage a continuum. But if we focus on a continuum, we find that strain involves a *surface* approaching or separating from another *surface*, and, in the limit of small size, a planar element approaching another planar element. If the change is to occur at constant material density, it is a *wafer* of material that needs to be removed; removing a particle is not the operation that does the job.

Of course, eventually, particles will reappear, through the equation $\mu^A = \mu_0^A + RT \ln X^A$; the existence of R depends on the existence of particles. But a theory of stress-driven deformation of a continuum does not require particles, even with stress-driven self-diffusion; coefficients for viscosity and self-diffusion are the only things required.

Lack of realism. Figures 11.5b and 11.6b are fanciful rather than realistic. Yet the argument advanced hinges upon features shown in those figures, and hence the validity can be questioned: does the lack of realism invalidate the argument? A general answer is offered in this section, and a more specific answer concerning the length L_0 in particular is offered in the next section. The general answer begins with the thought that any continuum is unrealistic; continuum models have advantages, but realism is not one of them. A continuum model should be mathematically tractable, and should imitate *in selected ways* the behavior of real materials that one wishes to elucidate. We wish to elucidate simultaneous self-diffusion and deformation; hence it is necessary that the continuum we discuss should have coefficients for these two behaviors. Beyond that, we should leave it as featureless as possible: we assume continuity, we assume the two coefficients exist, we assume linear behavior, conservation and constant density, but beyond these attributes, we assume nothing. We certainly do not need to assume that the continuum actually contains a remarkable set of tiny slots, that have length but no width and cross each other without intercommunicating. We do not assert that the continuum has these; we assert only that its behavior is *as if* it had them. In other words, we use two levels of unreality: the continuum is an unreal model of a real material, and the swarm of slots is an unreal model of the continuum's own unreal behavior. To turn things around and emphasize the positive, the continuum is an idealization that helps us think about certain behaviors in real materials, and equally the slots are an idealization that helps us think about certain behaviors in the postulated continuum. The next section adds specificness to this general statement.

Significance of the length L_0. The continuum in view has two responses to stress. In a homogeneous stress field, it can change shape at constant volume; a viscosity coefficient N is used to describe the responsiveness here. Or in an inhomogeneous stress field, there can be diffusive mass transfer as well (with local change of volume) and a diffusion coefficient K can be used to describe the extra diffusional response. The standard unit for N is Pa-sec; also, if the diffusive response is measured in cubic meters of material passing across a reference plane per square meter per second, under a driving gradient in pascals per meter, the unit for K is m^2/Pa-sec. Then $(NK)^{1/2}$ is a length that is as much a characteristic of the continuum in view as N or K.

In Figure 11.5, a swarm of arcs or quadrants was imagined, their length was named L_0, and it was claimed that if the material flows and diffuses, it behaves *as if* such arcs act as migration paths. It is shown in Chapter 12 that if a material has macroscopic properties N and K, the imagined arc-length L_0 needs to equal $\pi(NK)^{1/2}$, whereas to maximize the simplicity of the master equation (12.8), the length used to multiply the angle α needs to equal $2(NK)^{1/2}$. These different arithmetic multiples of $(NK)^{1/2}$ are less important than the question: Can a continuum have a characteristic length and still have the properties of an ideal continuum, or does identifying a characteristic length force us to abandon the continuum concept and imagine instead a material with some kind of microstructure? It will be proposed that we are *not* so forced. The history of this topic is rich in discussions that *begin* with some concept of microstructure (atoms, molecules, vacancies, dislocations, etc., as in Figures 11.1 and 11.2; see Appendix 12A) but we need to escape from this history. Statistical mechanics, including lattice dynamics as a subfield, is a proper field of enquiry that requires microstructure, but the thermodynamics of a continuum is different. In continuum theories, one may *choose* to postulate microstructure (and useful and interesting results emerge) but one is not *obliged* to do so. It seems to me useful to imagine a continuum with properties N and K, to calculate how it would behave in various circumstances, and to compare that behavior with real life. $(NK)^{1/2}$ certainly is a length expressible in nanometers, but in following the train of enquiry mentioned, one is not obliged to decorate the continuum with features at spacings such as $(NK)^{1/2}$ apart.

Summary and conclusions

A real viscous material in a homogeneous nonhydrostatic stress field changes shape at some rate that enables us to assign the material a viscosity N.

In an equivalent imaginary continuum, the change of shape involves removal of infinitesimal wafers of material from sites normal to a principal direction of shortening, and insertion of wafers of material at sites normal to the principal direction of elongation.

Behavior of the continuum is *as if* the wafers slid sideways in their own plane along cylindrical slots.

The driving agents for the change of shape are the differences among the principal stresses; in the special case of cylindrical constriction, the driving agent is the difference $\sigma_{\text{radial}} - \sigma_{\text{axial}}$ where σ is a normal-stress component. The shortening rate across a plane subject to σ_{radial} is $(\sigma_{\text{radial}} - \sigma_{\text{axial}})/6N$. Behavior is *as if* each wafer was driven along a slot by the gradient in the stress normal to the slot from σ_{radial} at one end to σ_{axial} at the other.

Turn now to a nonhomogeneous stress field where, in addition to change of shape, there is diffusive mass transfer from sites of higher compression to sites of lower compression: it is possible to devise a stress field as in Figure 11.6 such that successive points across a planar surface have the same sequence of normal-stress components as successive points across a cylindrical surface in the homogeneous constrictive field.

If the material responds to a gradational sequence of normal-stress states, it is not reasonable to suppose that the response will be different according as the normal-stress components act on elements that form a cylindrical surface or a planar surface. Thus the shortening rate across a plane subject to σ_{max} will be $(\sigma_{\text{max}} - \sigma_{\text{min}})/6N$ by diffusive mass transfer, for some small wavelength λ in the stress-field fluctuation (λ in the range nanometers to micrometers).

If the wavelength λ could be determined experimentally, it would be a direct indicator of the arc-length of an imaginary quarter-cylinder; then, whatever the mechanism is by which change of shape occurs in a homogeneous constrictive stress field, the actual mechanism and the imagined quarter-cylinder migration would be equivalent. In other words, any material that has a viscosity N Pa-sec and a self-diffusion coefficient K m^2/Pa-sec behaves *as if* it contained cylindrical slots of radius $2(NK)^{1/2}$.

In a material of just one component, shortening by diffusive mass transfer and shortening by change of shape at constant volume are additive processes. The net result takes the form

$$e_{\text{total}} = \text{d-m-transfer terms} + \text{change-of-shape terms} \qquad (11.2)$$

the terms having the forms $C_1(\partial^2\sigma_{zz}/\partial x^2)$ and $C_2(\partial^2\sigma_{zz}/\partial\theta^2)$. Here C_1 and C_2 are coefficients closely related to K and N.

The details of the coefficients are treated in Chapter 12. The essence of this chapter is to establish the pivotal role of the imagined quarter-cylinder migration paths:

1. An ideal continuum behaves as if it contained them.
2. To the extent that a real atomic material behaves like an ideal continuum, it too behaves as if it contained quarter-cylinder migration paths.
3. The radius of the paths that the atomic material *appears* to have (as regards macroscopic behavior) is related to the actual atomic mechanism by which change of shape occurs.

APPENDIX 11A: OTHER HARMONIC STRESS FIELDS

No new ideas are contained in this appendix; the intention is to increase familiarity with ideas already proposed by applying them in two situations that are slightly different from the one already shown in Figure 11.6 and eqs. (11.1). The first change is from the cylindrically symmetrical situation to a plane strain, retaining the proviso that the compression along x is uniform. Then the second change is to relax that restriction, and introduce a gradient in σ_{xx} that is equal in magnitude and opposite in direction to the gradient in σ_{zz}.

Plane strain

In diffusion-free materials, one can normally produce plane strain by making $\sigma_{yy} = \frac{1}{2}(\sigma_{xx} + \sigma_{zz})$. Where self-diffusion operates, the result is not exactly a plane strain, as the following analysis shows.

In place of the stress field specified by eqns. (11.1) we use

$$\sigma_{zz} = \tfrac{1}{2}[\sigma_1 + \sigma_3 + (\sigma_1 - \sigma_3)\cos 2\pi x/\lambda] \qquad (11A.1a)$$

$$\sigma_{yy} = \tfrac{1}{2}(\sigma_{zz} + \sigma_{xx}) \qquad (11A.1b)$$

$$\sigma_{xx} = \tfrac{1}{2}(\sigma_1 + \sigma_3) \qquad (11A.1c)$$

All the ideas discussed before remain unchanged, as far as they concern response to the gradient $\partial\sigma_{zz}/\partial x$ and the stress difference $(\sigma_{zz} - \sigma_{xx})$. A difference is that there is now an additional stress difference $(\sigma_{zz} - \sigma_{yy})$, equal to $\frac{1}{2}(\sigma_{zz} - \sigma_{xx})$. In consequence, the shortening rate along z by change of shape is larger than before by the factor $3/2$; it is now larger than the shortening rate due to diffusive mass transfer, instead of being equal to it.

We should also consider the shortening rate along y, which in the cylindrical case is equal to that along z. By changing to the condition $\sigma_{yy} = \frac{1}{2}(\sigma_{zz} + \sigma_{xx})$, we bring the effect of change of shape to zero, as in plane strain of a nondiffusing material. But shortening by diffusive loss does not become zero: $\partial^2\sigma_{yy}/\partial x^2$ diminishes just to half its former value, i.e., to $\frac{1}{2}\partial^2\sigma_{zz}/\partial x^2$. To eliminate this effect, it is necessary to make σ_{xx} vary along x as follows.

Reciprocal gradients

The second stress field of interest is

$$\sigma_{zz} = \tfrac{1}{2}[\sigma_1 + \sigma_3 + (\sigma_1 - \sigma_3)\cos 2\pi x/\lambda]$$
$$\sigma_{yy} = \tfrac{1}{2}(\sigma_{zz} + \sigma_{xx}) \qquad\qquad \text{as before}$$

(11A.2a,b)

$$\sigma_{xx} = \tfrac{1}{2}[\sigma_1 + \sigma_3 - (\sigma_1 - \sigma_3)\cos 2\pi x/\lambda]$$

(11A.2c)

See Figure 11.8; now both σ_{yy} and the mean stress are uniform along the x-direction.

Taking a simple view, one could suggest first that at site (i) there will be shortening along z and elongation along x at rates $(\sigma_1 - \sigma_3)/4N$ (N = viscosity), and that at site (ii) behavior will be the same, with signs reversed. Then if one supposes that diffusive mass transfer is controlled by mean stress, the rates just suggested must be the *whole* of the material's response, because no diffusive mass transfer would occur.

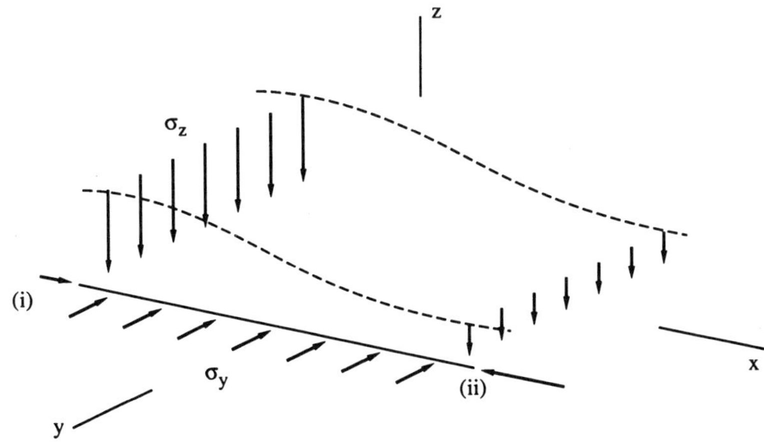

Figure 11.8 A variant on Figure 11.6 that gives plane strain in xz planes.

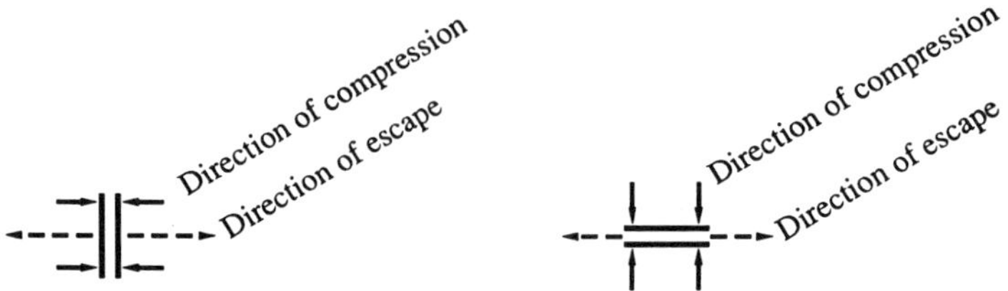

Figure 11.9 The two situations shown are fundamentally different; hence equal gradients $\partial\sigma_{xx}/\partial x$ and $\partial\sigma_{yy}/\partial x$ will not in general have similar consequences.

On the other hand, if diffusive mass transfer is controlled by normal-stress components, the suggestion would be that $\partial^2\sigma_{zz}/\partial x^2$ would drive shortening along z at site (i) and elongation along z at site (ii), while $\partial^2\sigma_{xx}/\partial x^2$ would drive shortening along x at site (ii) and elongation at site (i). The geometrical effects would be of the same type as the classical constant-volume change-of-shape effects, but additional to them.

We have here then, in concept, a rather clear observational comparison of the two theories. Let an initial experiment be performed using a homogeneous stress field with principal stresses σ_1 and σ_3; then a portion of the sample is kept in this stress state while the stress state nearby is changed to the one shown at site (ii). According to the mean-stress theory, behavior at site (i) will be wholly unaffected, but according to the normal-stress-component theory, the strain rates at site (i) will increase. In fact when $\lambda/2 = L_0$, the strain rate along z at site (i) should be just 5/3 times its magnitude in a homogeneous plane strain, or twice its magnitude in the cylindrically symmetrical case.

The experiment just described would be of great interest: the predicted doubling of strain rate could be sought, and any change in strain rate along x could also be examined: it would be of interest because, even when the derivatives of σ_{zz} and σ_{xx} are equal, one need not suppose the resulting diffusive mass transfer rates will be equal—the situations shown in Figure 11.9 being physically different.

As noted on page 93, it is conjectured that L_0 is of the order of 1 μm so that, even if the postulated stress field could be approximated, observing strain rates with sufficient local detail would be difficult. Here the purpose of the discussion is more to practice with the ideas in abstract form. Experimental work is touched on in Chapter 20.

APPENDIX 11B: GIBBSIAN ARGUMENT FOR MULTIVALUED POTENTIALS

The following argument supports the idea that a component's chemical potential at a point is multivalued. It is a free-standing argument independent

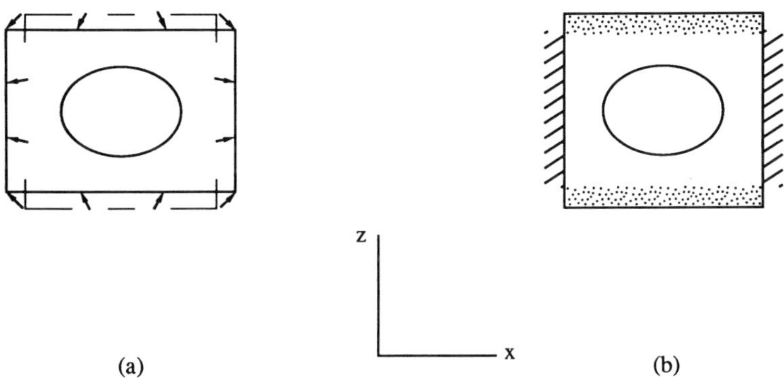

Figure 11.10 Two thermodynamic systems in which material strains homogeneously. In (a) the system boundaries move with the material, and forces on the boundaries perform work. In (b) the system boundaries remain stationary, but there is a flux of material inward across one pair of faces and outward across another pair. Stipple shows material added and diagonal lines show an equal amount of material lost.

of the content of Chapter 11; the chapter and this appendix can be taken in either order.

An extensive homogeneous sample contains a homogeneous stress field with principal values σ_x, σ_z, and $(\sigma_x + \sigma_z)/2$. The material is Newtonian and deforms in a steady manner with principal strain rates e and $-e$. Energy is dissipated at a rate $(\sigma_z - \sigma_x)e$ per second per unit volume, and energy is withdrawn from the sample at this rate uniformly throughout its volume by some unspecified process, so that the sample's temperature is also steady through time. We identify a portion in the interior of the sample that at some moment is a cube with linear dimension k. We ask where is the source of supply of energy to this cube, with power $k^3(\sigma_z - \sigma_x)e$.

Two answers can be given by considering the thermodynamics of the cube. In one approach we consider a thermodynamic system whose boundaries coincide with those of the cube and move with the material; in the other we consider a thermodynamic system whose boundaries coincide with the cube at some instant but remain fixed in space so that material flows across them (i.e., a Lagrangian element and an Eulerian element, respectively)—see Figure 11.10.

For the Lagrangian element, the source of energy is the work done on the boundaries of the system as they move—with power $\sigma_z k^2 \cdot ek$ on the pair of faces normal to z and $-\sigma_x k^2 \cdot ek$ on the pair normal to x (where energy is expended rather than absorbed); but what is the energy source for the thermodynamic system with stationary boundaries? The boundaries being stationary, no work is done on this system; the system is maintained in a state that is steady through time by fluxes of mass, inward across boundary planes normal to z and outward across bounding planes normal to x. Let these fluxes be f and $-f$ so that, per second, a mass fk^2 enters, which we designate δm. Associated with the entry of this mass across faces normal to z is a change in the free enthalpy or Gibbs free energy of the system, which we designate δG_z, and associated with the exit of an equal mass at faces normal to x there is a change in energy that we designate δG_x. If δG_z and δG_x differ from each other by $ek^3(\sigma_z - \sigma_x)$, it is easy to see that dissipation of energy within the cube can be steady with time, but if they do not differ by this amount, one is puzzled to see how the whole process can remain steady.

The idea that energy cannot simply appear out of nowhere seems to force us to believe that δG_z and δG_x do differ by the stated amount. We form two quantities, $\mu_z = \delta G_z/\delta m$ and $\mu_x = \delta G_x/\delta m$, and note that, if the entry and exit process were reversible, μ_z and μ_x would be chemical potentials as defined by Gibbs. That is to say, if crossing a boundary normal to z were an equilibrium process, the chemical potential associated with the process would be μ_z, and if crossing a boundary normal to x were an equilibrium process, the potential associated with the process would be μ_x. Then for the real (nonequilibrium) processes that occur at the boundaries, if we identify associated equilibrium processes for them, we *have to* identify processes for which the potentials differ by $\mu_z - \mu_x$; there *has to* be a difference in the associated equilibrium potentials of this amount if they are to be useful in describing this kind of behavior.

An essential part of the argument just offered is the distinction between the mass δm that enters across z-faces and the mass δm that leaves across x-faces. These masses do not "enter the thermodynamic system" in any abstract way: they enter at specified locations and have specified effects on the linear dimensions of the system, i.e., they exactly compensate for the internal straining. In other words, the thing introduced at a z-face during a short interval of time is a *wafer* of material of area k^2 and thickness $ke \cdot \delta t$. The concept of material arriving at a site or leaving in the form of a wafer is used again in the body of Chapter 11, especially in Figure 11.4. It is in fact an essential idea throughout Part II.

At this point it is convenient to make another distinction. The units of chemical potential are J/kg (or some similar unit) and a tendency exists to think of the joules of energy being dispersed through the kilogram, as one would if the topic were, for example, the hydrogen-atom content of the kilogram. If one slices off one-quarter of the kilogram, one expects (in a simple world) to slice off one-quarter of the hydrogen atoms contained, regardless of how the slicing is done, because the hydrogen content is a characteristic of the material, not a characteristic of the slicing process. But chemical potential is different: as defined, it is a ratio of the energy-change to the mass-change attending some *process*—it is a characteristic of the *process* rather than a characteristic of the material being handled. In the present discussion, the material being added is very similar to the material that leaves—in fact, indistinguishable from it in all intrinsic respects.

Nonetheless, the *process* of adding is *not* the same as the *process* of removal, and the chemical potential of the adding process is not the same as the chemical potential of the removal process. The superficial resemblance of joules per kilogram to hydrogen atoms per kilogram tends to mask this important difference.

12

DEFORMATION AND DIFFUSION: QUANTITATIVE RELATIONS

The purpose of this chapter is to put the ideas of Chapter 11 into quantitative form. The first step is to link L_0 to N and K; L_0 is the arc-length of the imaginary quarter-cylinders in Figure 11.5b, N is the material's viscosity (Pa-sec), and K is its coefficient for pressure-driven self-diffusion (m^2/Pa-sec).

Relation of L_0 to N and K

The point emphasized in Chapter 11 is that if two migration paths exist, one curved and one straight, but both having the same length and the same variation of normal-stress components along their length, migration will be equally vigorous along the two paths. Further, the shortening rates at the source-ends of the two paths will be equal. The procedure used to find the relation of L_0 to $(NK)^{1/2}$ is to write the two shortening rates and equate them.

Change of shape. A relevant equation from Chapter 7 is, for any principal directions 1 and 3,

$$\sigma_1 - \sigma_3 = 2N(e_1 - e_3) \tag{12.1}$$

In the situation of Figure 11.6 and eqns. (11.1), let z be direction 1 and x be direction 3. To conserve volume and to conserve cylindrical symmetry, we need $e_{xx} = -2e_{zz}$ so that $(e_1 - e_3)$ or $(e_{zz} - e_{xx})$ becomes $3e_{zz}$, and $e_{zz} = (\sigma_1 - \sigma_3)/6N$.

Diffusive mass transfer. The central idea is expressed in eqns. (10.7) and (10.8). In a situation where variation in the potential μ arises *only* from variation in compressive stress, (10.7a) becomes

$$\frac{\partial \varepsilon_{11}}{\partial t} \quad \text{or} \quad \frac{k}{3}\nabla^2\mu_1 = \frac{kV}{3}\nabla^2\sigma_1 \tag{12.2}$$

Here the product kV corresponds to K in the main text, defined by $\partial v/(v\cdot\partial t) = K\,\nabla^2 P$ with units m^2/Pa-sec. For the harmonic variation in eqn. (11.1a),

$$\frac{\partial^2\sigma_{zz}}{\partial x^2} = -\frac{2\pi^2}{\lambda^2}(\sigma_1 - \sigma_3)\cos 2\pi x/\lambda \tag{12.3}$$

100

so that, at the source point $x = 0$,

$$e_{zz} = -\frac{K}{3}\frac{2\pi^2}{\lambda^2}(\sigma_1 - \sigma_3) \tag{12.4}$$

For comparison with the change-of-shape result, we need to take e as positive in shortening rather than elongation, so the negative sign is dropped. Then equating the two expressions for e_{zz} gives

$$\lambda^2 = 4NK\pi^2, \qquad L_0^2 = NK\pi^2, \qquad L_0 = \pi(NK)^{1/2} \tag{12.5}$$

Also the radius of the curved path is $2(NK)^{1/2}$.

Change of shape and diffusive mass transfer combined

The material's characteristic length L_0 can be used to put ideas from Chapter 11 into diagrammatic form; see Figure 12.1. Diagram (a) shows the variation of the normal-stress component along the curved migration path from Figure 11.5, and diagram (b) shows the corresponding variation along the planar path from Figure 11.6. Diagrams (d) and (e) have two horizontal axes, one for the position-variable x and the other for a variable that changes continuously along the curved path. For the latter variable, one might use the angle α shown in diagram (c) or the length $R\alpha = (2/\pi)L_0\alpha$. Either way, the diagram that represents the spatially uniform constrictive stress field is diagram (d), and the diagram that represents the harmonically varying field is diagram (e). The useful property of these diagrams is that they enable one to *see* a representation of the two variations in normal stress that lead to shortening along z.

Consider first the spatially uniform situation, diagram (d). Any normal-stress component in the profile, diagram (a), is given by

$$\tfrac{1}{2}[\sigma_1 + \sigma_3 + (\sigma_1 - \sigma_3)\cos 2\alpha]$$

so that $\partial^2\sigma_\alpha/\partial\alpha^2 = -2(\sigma_1 - \sigma_3)\cos 2\alpha$; or if we use $R\alpha$ as variable, $\partial^2\sigma_\alpha/\partial(R\alpha)^2 = -(2/R^2)(\sigma_1 - \sigma_3)\cos 2\alpha$. In particular, the maximum value, at $\alpha = 0$, is $2(\sigma_1 - \sigma_3)/R^2$.

For the spatially varying field, diagram (e), the maximum value of $-\partial^2\sigma_{zz}/\partial x^2$ comes directly from eqn. (12.3) and is $2\pi^2(\sigma_1 - \sigma_3)/\lambda^2$. $R = 2L_0/\pi$ and $\lambda = 2L_0$, so the two second derivatives are equal as intended and as shown in the diagram.

The two equal contributions to the strain rate e_{zz} can be shown thus:

$$e_{zz}(\text{total}) = \frac{1}{12N}\frac{\partial^2\sigma_{zz}}{\partial\alpha^2} + \frac{K}{3}\frac{\partial^2\sigma_{zz}}{\partial x^2} \tag{12.6a}$$

Here the second right-hand term comes from eqn. (12.2) and the first comes from eqn. (8.11); $\partial^2\sigma_{zz}/\partial\alpha^2$ means $\partial^2\sigma_\alpha/\partial\alpha^2$ taken at $\alpha = 0$, where $\sigma_\alpha = \sigma_{zz}$.

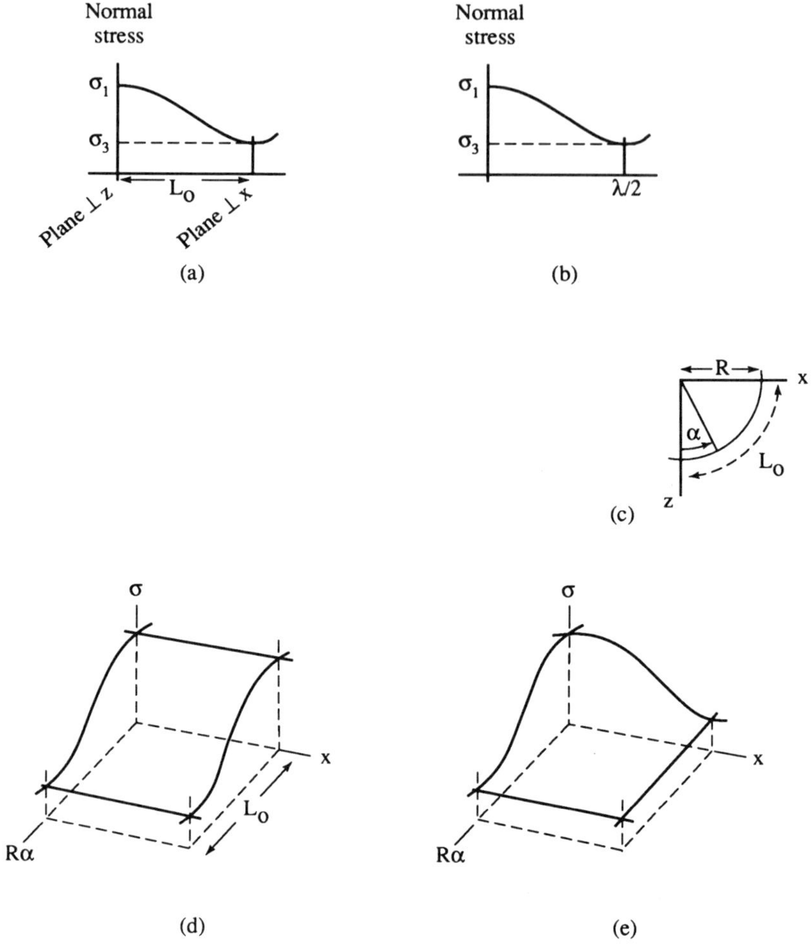

Figure 12.1 Variation of compressive stress σ with position along x and with orientation.

Again, the resemblance between Figures 11.5 and 11.6 suggests replacing α itself by the length $R\alpha$:

$$e_{zz}(\text{total}) = \frac{R^2}{12N}\frac{\partial^2\sigma_{zz}}{\partial(R\alpha)^2} + \frac{K}{3}\frac{\partial^2\sigma_{zz}}{\partial x^2} \tag{12.6b}$$

But $R = 2L_0/\pi = 2(NK)^{1/2}$, so that

$$e_{zz}(\text{total}) = \frac{K}{3}\left(\frac{\partial^2\sigma_{zz}}{\partial(R\alpha)^2} + \frac{\partial^2\sigma_{zz}}{\partial x^2}\right) \tag{12.6c}$$

The result (12.6c) is a formal statement. It is not new: the idea it embodies was expressed in Chapter 11 by saying, for the planar and the curved migration paths, that it is not reasonable to suppose that migration behavior will be different if the normal-stress variations along the two paths are the same. That idea led to the *selection* of a length L_0 and radius R so that the one coefficient K would describe both change of shape and diffusive mass transfer. But the equation incorporates eqns. (10.7): the choice of R gives the equation its simple form, but it is the discussion of the two equivalent migration paths that forces use of the term $\partial^2 \sigma_{zz}/\partial x^2$. It is the material's inability to behave differently on the two paths that makes σ_{zz}, not the mean stress, the quantity whose variation along x we need to follow.

Generalizations. Consider a general stress field, both nonhydrostatic and spatially nonuniform. For any direction $\mathbf{n}$ we can consider two orthogonal variations of the normal-stress component σ_{nn} with orientation, as in Figure 8.4. As shown in Appendix 8A, the sum $\partial^2 \sigma_n/\partial\theta_1^2 + \partial^2 \sigma_n/\partial\theta_2^2$ is a function only of $\mathbf{n}$; it is invariant with choice of the orthogonal planes that contain θ_1 and θ_2, in the same way that the Laplacian is invariant with choice of x_1, x_2, and x_3. And, again from Appendix 8A, the relation of shortening rate to $\partial^2 \sigma_n/\partial\theta^2$ is general for any direction; it is not specific for the direction of *maximum* compressive stress σ_{zz} as discussed above. Thus equation (12.6c) generalizes to

$$e_{nn} = \frac{K}{3}\left[\frac{\partial^2}{\partial(R\alpha_1)^2} + \frac{\partial^2}{\partial(R\alpha_2)^2} + \frac{\partial^2}{\partial x_l^2} + \frac{\partial^2}{\partial x_m^2} + f\frac{\partial^2}{\partial x_n^2}\right]\sigma_{nn} \qquad (12.7)$$

Here directions l and m are any two directions normal to $\mathbf{n}$. The factor f is inserted for reasons shown in Figure 11.9; it lies between 0 and 1 and is discussed in more detail on page 104.

To continue we use arguments from Chapter 9: in the same way that eqn. (8.11) generates (9.4), we can again convert from σ_{nn} as the driving variable to the associated equilibrium chemical potential μ_n. The factor needed is V, the material's unit volume; the material coefficient becomes $K/3V$, and introducing the symbol D for this coefficient gives

$$e_{nn} = D\left(\frac{\partial^2}{\partial(R\alpha_1)^2} + \frac{\partial^2}{\partial(R\alpha_2)^2} + \frac{\partial^2}{\partial x_l^2} + \frac{\partial^2}{\partial x_m^2} + f\frac{\partial^2}{\partial x_n^2}\right)\mu_{nn} \qquad (12.8)$$

$D = K/3V$; K is defined by $\partial v/(v\,\partial t) = K\nabla^2 p$; V is the material's unit volume; $R = 2(NK)^{1/2}$. Units are:

e	$\sec^{-1}$
D	kg/(Pa-m-sec)
K	m^2/(Pa-sec)
V	m^3/kg
N	Pa-sec
R	m
μ	J/kg

Discussion. In eqn. (12.8), the factors R, μ, and f all deserve comment.

Regarding R, we note that the idea of a material having a characteristic length in the range from nanometers to micrometers has been entertained for nearly 100 years. Appendix 12A reviews variants on this basic idea that have been proposed, mostly on the basis of one version or another of the behavior of atoms. The result, among many workers, has been unanimity as regards the factor $(NK)^{1/2}$ and diversity of opinion about a suitable arithmetic multiplier in front. By contrast, the approach taken here emphasizes the *absence* of atoms from an ideal continuum; the factor of 2 arises strictly from the mathematics. It is therefore regarded as being in some sense a more fundamental number than any number derived from imagined behavior of rigid spheres, etc.

Regarding μ, it is proper to be cautious about the conversion just made from eqn. (12.7) to (12.8). As long as the circumstances are such that variation in μ is present *only* as a consequence of variation in σ, it is clear that (12.7) implies (12.8). But the whole value of μ as a predictor of behavior is its umbrella quality: in other areas of science, the usefulness of μ arises from the fact that an increase due to change of temperature or an increase due to change of pressure lead to similar consequences—a gradient in μ has similar effects regardless of the reason for the gradient being present. This experience leads toward the idea that eqn. (12.8) may continue to describe behavior where μ varies for other reasons in addition to variation in σ. In particular, we wish to treat a material of nonuniform composition. In such a material, $\partial \mu / \partial x$ has a part related to $\partial \sigma / \partial x$ and also a part related to $\partial(\text{composition})/\partial x$. At this point, it is merely *suggested* that eqn. (12.8) will continue to be useful; the topic is pursued in detail in Chapter 14.

Regarding the factor f, again we make here only two preliminary remarks. First, it is possible to pass by a *continuous* series of changes from a robust glass through various fluids of increasing mobility into the supercritical region, that is, into a gas. At the glass end, we wish to apply the equations of continuum mechanics and use the assertion made that a linear strain involves a *surface* approaching another *surface*, which leads in turn to the entry of the two $R\alpha$-terms into (12.8). At the gas end, the concept of a *surface* within the continuum is less useful, the $R\alpha$-terms are likely to be less important and f tends toward 1; that is, the right-hand side tends toward the classical Laplacian form. The idea emerges that where neighboring particles in a material have strongly correlated motions (e.g., a glass), the $R\alpha$ terms are important and f may be less than 1; but where neighboring particles are not correlated (e.g., a gas) a classical Laplacian approach serves. (Presumably there is a difficult region in between where neither of the simple end-member approaches will work.)

The second remark is that, for the purposes of the present text, we emphasize stiff continua and the importance of the $R\alpha$ terms. But the suggestion just made, that where particles are independent f tends toward 1, is relevant as follows: when μ varies with σ and with concentration, $\partial^2 \mu / \partial x^2$ has two parts $(A\ \partial^2 \sigma / \partial x^2)$ and $(B\ \partial^2 (\text{conc})/\partial x^2)$; we may wish to put $f = 1$

for the second of these even while emphasizing that $f \neq 1$ for the first. That procedure would allow for the fact that when a material changes shape in response to stresses, all the constituent particles move together in a highly correlated way, whereas when the material responds to nonuniform concentration, atoms display more the independent behavior characteristic of the gaseous state. Again, Chapter 14 will permit these ideas to be illustrated by an example.

APPENDIX 12A: A MATERIAL'S CHARACTERISTIC LENGTH

The purpose of this appendix is to comment briefly on the idea that a material has a characteristic length. In particular it recounts a few notable occurrences of this idea, taking as starting point a paper by Albert Einstein (1905). But before starting on history, it is as well to emphasize a difference between the material of this appendix and the main text.

The essential difference is that in this appendix the geometry of atoms and interatomic distances is discussed, whereas the policy in the main text is to avoid such details. The objective of the main text is to put together the simplest possible equations that describe a material's creep and chemical change; for this purpose we imagine an ideal continuum, whose properties are the same at every point or change in a mathematically continuous way. We assume that such a continuum can creep and self-diffuse and, as noted, just these two macroscopic behaviors create the concept of a length $2(NK)^{1/2}$. Most people find that to imagine self-diffusion occurring they have to imagine that, in some form and on some scale, the continuum is divided into particles, but in the main text as little attention as possible is given to that aspect; certainly, the idea of a length $2(NK)^{1/2}$ has no relation whatever to any particle diameter or interatomic distance. By contrast, Einstein and subsequent workers begin by postulating atoms of definite sizes, and if they end by writing an equation relating N and K to a length L, the length is related to the atoms' radii. It is interesting to see how what may be called *ball models* do two things: on one hand, they lead toward a general idea of how materials behave by providing particular instances; but on the other hand, they prevent people from forming the general ideas—thinking about balls can be a distraction and a trap instead of being a stimulus toward generalizations.

The Stokes–Einstein theory

Stokes wrote equations for the movement of a rigid sphere through an ideal continuous viscous fluid. Einstein (1905) subsequently considered the movement of atoms of species A through a mass of atoms of species B. Using the Stokes equations, he produced results for the condition where atoms of B are sufficiently small in comparison with atoms of A to act as a featureless fluid host. The result is the relation

$$D_A^i = \frac{kT}{6\pi r_A N_B}$$

Here k = the Boltzmann constant, r_A is the radius of an atom of species A, and D_A^i is the diffusion coefficient for movement of A driven by a concentration gradient. The corresponding relation for K_A, the pressure-driven coefficient, is gained by multiplying by V_A/RT,

$$K_A = \frac{kV_A}{R6\pi r_A N_B}$$

and here kV_A/R is the volume occupied by just one atom of A. In isolation this would be $4\pi r_A^3/3$ or about $4.2r_A^3$, but even when spheres are close-packed the effective volume occupied by each is larger, about $5.6r_A^3$. The product $K_A N_B$ then comes to $0.3r_A^2$. If we go on and use the formula

$$\text{Characteristic length} = 2(KN)^{1/2}$$

the length calculated is $1.1r_A$, but this is of course not the characteristic length of any single material, because we have multiplied the diffusivity of species A by the viscosity of species B. If anything, it is a characteristic length for the diffusing system as a whole, or for the mechanism by which A moves through B.

Self-diffusion: the Eyring theory

The situation where the migrating molecules are the same size as the molecules among which they move, as for example in a one-component liquid, was described by Glasstone, Laidler, and Eyring (1941). The distance or length used in their theory is not the radius of any particle, but the distance a molecule moves when it jumps. For this distance to be well defined, one needs to have in view a material with a sufficient degree of compactness and order. Their result comparable with the Stokes–Einstein equation is

$$D^i = \left(\frac{\lambda_1}{\lambda_2\lambda_3}\right)\frac{kT}{N}$$

where λ_1, λ_2, and λ_3 are jump distances in three directions, D^i being evaluated along direction 1. Again we convert from D^i to K using the factor V_A/RT:

$$K_A = \frac{\lambda_1}{\lambda_2\lambda_3}\frac{kV_A}{RN} = \frac{\lambda_1}{\lambda_2\lambda_3}\frac{(\text{molecule volume})}{N}$$

The authors suggest that, at least approximately, the molecule's volume equals $\lambda_1\lambda_2\lambda_3$, so that $KN \cong \lambda_1^2$ and the characteristic length $2(KN)^{1/2} \cong 2\lambda_1$. Using measured values of K and N for water gives $\lambda_1 = 0.14\,\text{nm}$ and characteristic length $= 0.28$ nm, which are close to the radius and diameter of a water molecule's equivalent sphere.

Beside these two examples, there is the general fact that if a pressure-driven diffusivity expressed in m^2-Pa-sec is multiplied by a viscosity expressed in Pa-sec, the product has dimensions $(length)^2$. Thus for several decades, the ideas have been accepted that

1. by combining viscosity and diffusivity information, abstract quantities with dimension $(length)^2$ or $(length)$ can be derived, and also
2. in certain circumstances, an explicit model of a real material can be constructed, with particle diameter, particle separation, jump distance, vacancy dimensions, etc., all expressed in nanometers; using such a model plus assumptions about jump frequency, relations between diffusivity, viscosity and system dimensions can be fully written out.

But for any material that cannot be well described as an aggregate of globular molecules, there has been uncertainty: if a length is calculated as in (1), what is that the length of? What in the real world has that length?

To recapitulate ideas from the main text, here the following are proposed.

1. A homogeneous change of shape involves rearrangement of material.
2. The rearrangement that actually occurs is *as if* material migrated along quadrant arcs, the arc length or arc radius being a measure of the ease or difficulty of the change.
3. The length l calculated by multiplying out the quantity $2(NK)^{1/2}$ is the radius of an imaginary quadrant that is equivalent to the real process, whatever that may be, as long as one process by itself gives both the change of shape expressed in the viscosity N and the migration expressed in the diffusivity K.
4. Where the real process is some simple atomic movement, l will be simply related to the atomic dimensions. Where the real process is more complicated, l will continue to indicate the nature of that process by showing its effective or equivalent path length, even though no simple relation to atomic dimensions remains.

The present proposals thus extend earlier ideas without conflicting with them.

APPENDIX 12B: HARMONICALLY VARYING STRESS FIELD WITH NONSPECIAL WAVELENGTH

In the main part of Chapter 12, particular attention is paid to the condition where shortening by change of shape and shortening by diffusive mass transfer are equal in magnitude; the different forms of eqn. (12.6) all describe this condition and emphasize the crest-to-trough separation L_0 at which it is obtained. But for later applications we need more general equations, and it is convenient to assemble them here.

Equation (12.6a) can be rewritten

$$\text{shortening} \qquad e_{zz}(\text{total}) = \frac{(\sigma_{zz} - \sigma_{xx})}{6N} - \frac{K}{3}\frac{\partial^2 \sigma_{zz}}{\partial x^2} \qquad (12\text{B}.1)$$

If the transverse stress σ_{zz} varies harmonically along x but not with the special wavelength $2L_0$ or the special mean value σ_{xx}, we need the general form

$$\sigma_{zz} = \sigma_{xx} + S_0 + s \sin jx \qquad (12\text{B}.2)$$

with wavenumber j and wavelength $2\pi/j$. Then

$$e_{zz} = \frac{S_0 + s \sin jx}{6N} + \frac{Kj^2}{3} s \sin jx \qquad (12\text{B}.3\text{a})$$

This can be written in various ways to emphasize different aspects:

$$e_{zz} = \frac{S_0}{6N} + s \sin jx\left(\frac{1}{6N} + \frac{Kj^2}{3}\right) \qquad (12\text{B}.3\text{b})$$

$$e_{zz} = \frac{S_0}{6N} + \frac{s \sin jx}{6N}\left(1 + \frac{j^2 R^2}{2}\right) \qquad (12\text{B}.3\text{c})$$

$$e_{zz} = \frac{S_0}{6N} + s \sin jx \,\frac{K}{3}\left(\frac{2}{R^2} + j^2\right) \qquad (12\text{B}.3\text{d})$$

Points to note are:

1. The inverse relation of the viscosity N and the diffusivity coefficient K. Materials that deform by some recognizable mechanism have a characteristic length R for that mechanism that is fundamentally geometrical and thus insensitive to small changes in, for example, temperature; then heating or other change that increases K is likely to diminish N by about the same factor.
2. The material is more responsive to the fluctuation s than to the uniform mean value S_0. Because response to s includes self-diffusion along x as well as change of shape, the material behaves with a smaller apparent viscosity or greater apparent mobility, by the factor $(1 + j^2 R^2/2)$.
3. The response to the fluctuation s increases with the wavenumber j, but the effect is small until j^2 approaches $1/R^2$, i.e., until the wavelength diminishes to the order of magnitude $2\pi R$, in the micrometer to nanometer range. It is on this same small scale that chemical fluctuations interact with the material's deformation behavior, as discussed in Chapters 14–16.

Equation (12B.1) describes the response e_{zz} if a prescribed fluctuating stress σ_{zz} is imposed, but it can be used to answer a related question: if a material of fluctuating viscosity is subjected to a uniform constriction, what stress field is produced? We abbreviate $s(1 + j^2 R^2/2)$ as s', so that eqn. (12B.3c) becomes

$$e_{zz} = \frac{S_0}{6N}\left(1 + \frac{s'}{S_0}\sin jx\right) \tag{12B.3e}$$

If the viscosity N itself fluctuates to a small extent according to $N = N_0[1 + (n/N_0)\sin jx]$, clearly e_{zz} will not vary with x if $s'/S_0 = n/N_0$. In other words, if a fluctuation n is prescribed and the material is subjected to uniform constriction, a stress fluctuation s will result; when j is small, s is adequately approximated by $6ne_{zz}$ (self-diffusion negligible), but for large values of j the complete expression $6ne_{zz}/(1 + j^2 R^2/2)$ is needed; it leads to smaller magnitudes for s than the approximate form.

III

APPLICATION: MOVEMENTS ALONG ONE DIRECTION

The objective is to use ideas from Chapter 12 in some possible physical situations; especially, we wish to see whether eqn. (12.8) leads to predictions that we can understand and perhaps believe. The situations discussed continue to be highly idealized; we do not yet reach *practical* applications.

An example is as follows: suppose two phases meet at a planar interface and have some chemical component in common; then in equilibrium that component will be partitioned between the two phases in such a way that it has the same chemical potential in both. Now suppose a uniform deformation is imposed on the entire sample, including the interface: in what way will the equilibrium be disturbed, and what chemical changes will begin to run?

The approach to this question is shown in the figure. Chapter 13 treats strain of a sample that contains an interface but no variation in chemistry

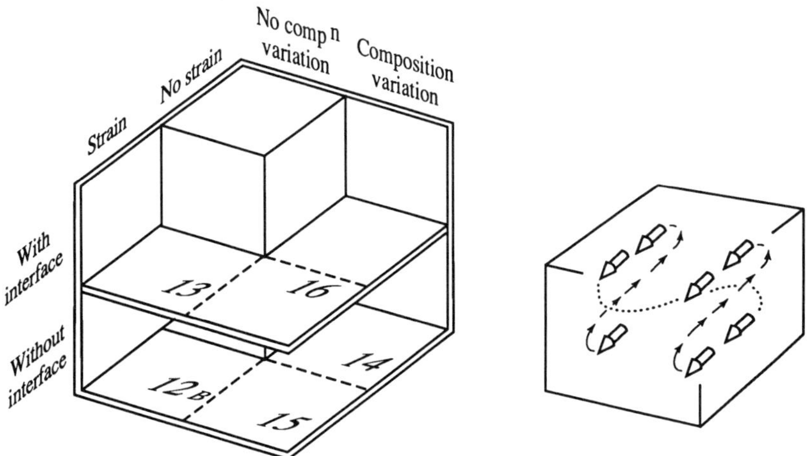

A diagram of possible conditions for a material sample, and the sequence of chapters in which the conditions are discussed.

(two polymorphs). Then Chapter 14 treats a sample with variation in chemistry but with no interface or external strain. Chapter 15 treats a sample with chemical variation and an imposed strain, but still no interface. Chapter 16 brings together ideas from Chapters 13 and 15, and addresses the question in the previous paragraph.

13
TWO PHASES AND ONE COMPONENT

It is at interfaces that the links between deformation and diffusive transfer are most noticeable. In general, stresses in a material vary only gradually from point to point, but at an interface—that is, a phase boundary—the variation can be abrupt. Figure 13.1 shows two stress profiles: if we take the interface as the yz plane of a coordinate system, the normal stress σ_{xx} is necessarily the same on both sides; but the normal stress on an xy plane, for example, namely σ_{zz}, is *not* necessarily the same. In fact the two

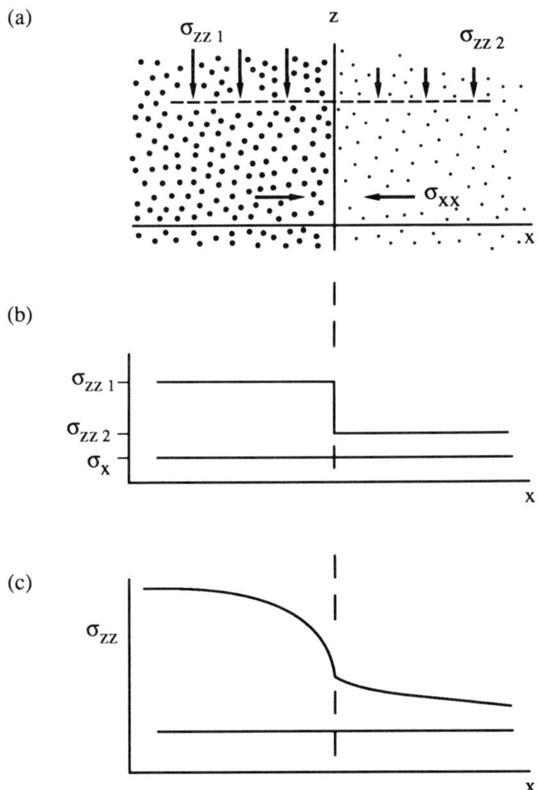

Figure 13.1 An interface between two phases that differ in viscosity. If the interface is coherent, the stress states on either side must be different.

magnitudes σ_{zz1} and σ_{zz2} are apt to be noticeably different. Assume for the moment that the interface is *coherent*, i.e., the two phases adhere so that any change of dimension on side 1 is exactly matched by an equal change on side 2. Then if one material is ten times stiffer than the other, either in elastic modulus or viscosity, we need

$$\sigma_{zz1} - \sigma_{xx} = 10(\sigma_{zz2} - \sigma_{xx})$$

Magnitudes might be $\sigma_{xx} = 0.1$, $\sigma_{zz2} = 1.1$, $\sigma_{zz1} = 10.1$ MPa, for example. For the purpose of designing machine parts, one can imagine a simple stress profile as in Figure 13.1b, with a vertical step or infinite gradient $\partial\sigma_{zz}/\partial x$; but if one is interested in chemical details at the interface, it is prudent to consider the possibility that the profile may actually be as in Figure 13.1c—with a strong but not infinite gradient. As will be shown later in this chapter, the width over which the main stress-drop extends is measured in nanometers or perhaps micrometers.

Initial conditions and assumptions

As in Figure 13.1a, the interface is taken as the yz plane. The material on the left, material 1, has properties N_1, K_1, R_1 (viscosity, diffusivity, and characteristic length, as in Chapter 12) and material 2 similarly. To be realistic one would assign the materials different unit volumes V_1 and V_2, but the effects of interest can be explored more directly if we assume initially that $V_1 = V_2$. One advantage of this assumption is that even if some material changes phase, or travels across the interface, or the interface migrates through the material (three ways of saying the same thing), material remote from the interface is not disturbed. We can therefore imagine a coordinate system pinned to this remote material, but can envisage material close to the interface possibly diffusing with respect to such a system.

As regards deformation, let all lines in the material that are initially straight and parallel to x remain so, and let them approach each other by a small cylindrically symmetrical constriction, elongating along x. Let strain rates be designated e_{xx}, etc.: then $e_{yy} = e_{zz} = $ a fixed value e_0, uniform everywhere, and at remote points (unaffected by diffusion) $e_{xx} = -2e_0$. But close to the boundary, diffusion effects allow e_{xx} to differ from $-2e_0$, even though we continue with $e_{yy} = e_{zz} = e_0$.

The situation is illustrated in Figure 13.2: a uniform set of elements is identified in the initial state; then we expect to have to *deform* these at constant area by different amounts as in diagram (b); some set of different amounts such as these are sufficient, because diffusive loss by migration across the interface will complete the reduction in height of each element by the uniform factor e_0. The problem is to find a profile for σ_{zz} like the one sketched in Figure 13.1c that drives the two processes simultaneously— diffusive loss and change of shape—in such amounts that the *total* vertical

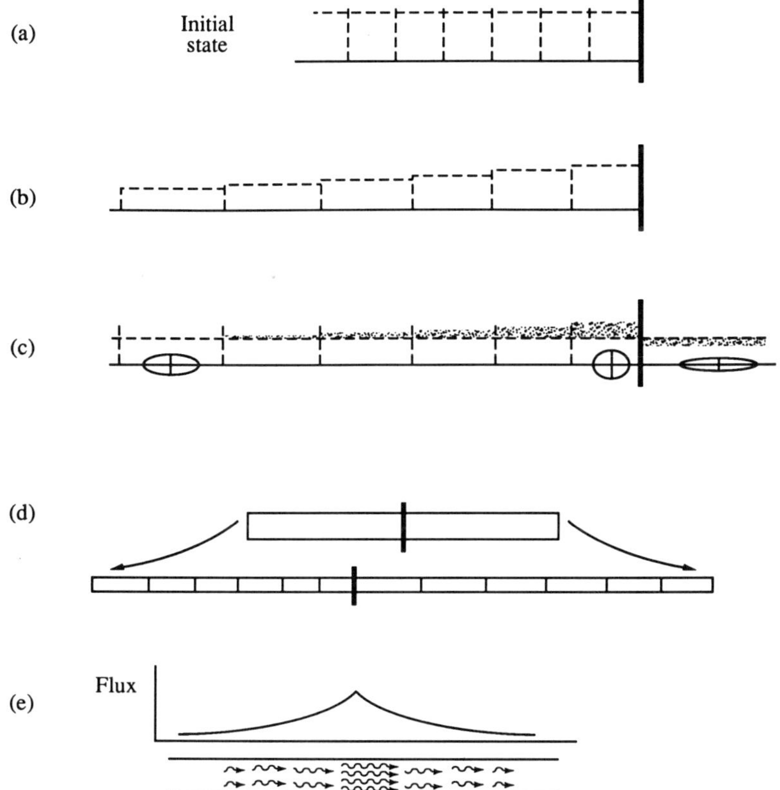

Figure 13.2 Deformation and diffusion at an interface. (a) The initial state, with a uniform set of elements identified. (b) The same elements in a later deformed state. (c) Stipple shows the amount of each element lost by self-diffusion (or *gained*, on the right-hand side of the interface). (d) The overall effect is a displacement of the interface. (e) The diffusive flux is a maximum at the interface.

shortening is the same at all points along x. The fact that change of shape is accentuated far from the interface and is milder close to the interface is indicated by the ellipses in Figure 13.2c.

On the other side of the interface, the low-stress side or downstream side, one can imagine comparable effects; the difference is that, on this side, material is *added* to elements close to the interface so that the change of shape produced by σ_{zz2} has to be *greater* for elements close to the interface than for remote points. The kind of profile needed for σ_{zz2} is shown in Figure 13.1c and the manner in which elements change dimension is shown in Figure 13.2d. As already suggested, the behavior of elements far from the interface is just as if a shortening e_0 and elongation $-2e_0$ were uniform throughout the whole sample; see the left-hand and right-hand ends of

Figure 13.2d. But near the interface flattening is less in material 1 and greater in material 2, so that the interface itself moves to the left with respect to the remote elements. In fact, rather obviously, the volume of space that the interface sweeps through equals the volume of material that crosses the interface by diffusion. The varying intensity of the diffusive flux must be somewhat as indicated in Figure 13.2e.

(In discussing the downstream side, an assumption is made about the manner in which material crosses the interface—that in leaving phase 1 in the form of wafers normal to z it also arrives in phase 2 as wafers normal to z. The possibility that diffusing material adds itself to phase 2 in the form of wafers normal to x also deserves consideration; but to proceed with the simplest possible analysis, the first assumption is retained. It amounts to assuming that the bond or weld at the interface is no weaker than the main mass of either material.)

Lastly, among initial conditions or assumptions, we consider what is known about stresses. Again, remote points are easy to discuss: to produce the radial shortening rate e_0, we need a stress difference $\sigma_{zz} - \sigma_{xx}$ of $6N_1e_0$ in material 1 and $6N_2e_0$ in material 2. If the diffusive movements have negligible effect on σ_{xx}, this stress component must be uniform throughout the material. Then at the remote ends, σ_{zz} has magnitudes $\sigma_{xx} + 6N_1e_0$ and $\sigma_{xx} + 6N_2e_0$.

At the interface, we do not know either σ_{zz1} or σ_{zz2} but it is reasonable to assume that they are equal. We already assumed that bonding across the interface was as strong and continuous as on either side, i.e., that the interface has no special weakness, and correspondingly we now assume that it is no kind of *barrier* either. If the interface were a barrier, then, for material to cross, one might ask about a stress difference or potential difference needed to drive diffusing material through the barrier at the needed rate; but if we assume a perfectly well-bonded interface, no stress difference or stress jump is needed and $\sigma_{zz1} = \sigma_{zz2}$.

We are now ready to use eqn. (12.7) to find the exact shape of the profile of σ_{zz}, and thence all other details of the processes occurring.

Solution

Equation (12.7) applies to any direction **n**. If we apply the equation specifically to the z-direction, it states

$$e_{zz} = \frac{K}{3}\left[\frac{\partial^2}{\partial(R\alpha_1)^2} + \frac{\partial^2}{\partial(R\alpha_2)^2} + \frac{\partial^2}{\partial x^2} + \frac{\partial^2}{\partial y^2} + f\frac{\partial^2}{\partial z^2}\right]\sigma_{zz} \qquad (13.1)$$

The constrictive stress is uniform along y and z—it is only along x that σ_{zz} varies; hence the fourth and fifth terms are zero. Also, let α_2 lie in the yz plane; then because of the cylindrical symmetry, the second term is zero also.

In fact, quite properly, we get back to eqn. (12.6):

$$e_0 = \frac{K}{3}\left[\frac{\partial^2\sigma_{zz}}{\partial(R\alpha_1)^2} + \frac{\partial^2\sigma_{zz}}{\partial x^2}\right] = \frac{1}{12N}\frac{\partial^2\sigma_{zz}}{\partial\alpha_1^2} + \frac{K}{3}\frac{\partial^2\sigma_{zz}}{\partial x^2} \qquad (13.2)$$

If α_2 lies in the yz plane, then the remaining angle, α_1, lies in the xz plane and $(\partial^2\sigma/\partial\alpha_1^2)_{zz} = -2(\sigma_{zz} - \sigma_{xx})$ as on page 101. Thus,

$$e_0 = -\frac{(\sigma_{zz} - \sigma_{xx})}{6N} + \frac{K}{3}\frac{\partial^2\sigma_{zz}}{\partial x^2} \qquad (13.3)$$

The uniform stress component σ_{xx} has no direct effects; it is the difference $\sigma_{zz} - \sigma_{xx}$ that produces results. We therefore name this difference S and write

$$e_0 = -\frac{S}{6N} + \frac{K}{3}\frac{\partial^2 S}{\partial x^2} \qquad (13.4a)$$

or

$$\frac{\partial^2 S}{\partial x^2} - \left(\frac{1}{2NK}\right)S = \frac{3e_0}{K} \qquad (13.4b)$$

or

$$-\frac{\partial^2 S}{\partial x^2} = \left(\frac{1}{2NK}\right)(S_{\text{remote}} - S) \qquad (13.4c)$$

The last of these equations is what we need; it expresses in exact form the ideas already sketched in Figures 13.1 and 13.2: at the far left-hand end, S approaches S_{remote} and $\partial^2 S/\partial x^2$ approaches zero; but as S diminishes toward the interface, the right-hand side becomes nonzero, and the profile needs to become curved. The curvature has to be such that for any local element, the diffusive loss—more flowing out on the right than flows in on the left—exactly makes up for the change-of-shape effect $S/6N$ no longer being as large as e_0. (The link between $\partial^2 S/\partial x^2$ and the local rate of loss of volume is illustrated in Figure 8.2 and the accompanying eqn. (8.6).)

A suitable profile for S is an exponential curve. Specifically, let

$$S_{\text{remote}} - S = Ae^{x/B} \qquad (13.5a)$$

Then $(Ae^{x/B})/B^2 = (Ae^{x/B})/2NK$, giving

$$B = (2NK)^{1/2} \qquad (13.5b)$$

Both A and B are shown in Figure 13.3. The factor A is the stress difference $(S_{\text{remote}} - S)$ evaluated at the interface, and the factor B is the distance from the interface where $(S_{\text{remote}} - S)$ has diminished to $1/e$ times its interface value.

The solution just given fits the left-hand material where x goes away to large negative values, so the A, B, N, and K discussed can all be given suffix (1). For the right-hand material, we need $S - S_{\text{remote}} = A_2 e^{-x/B_2}$.

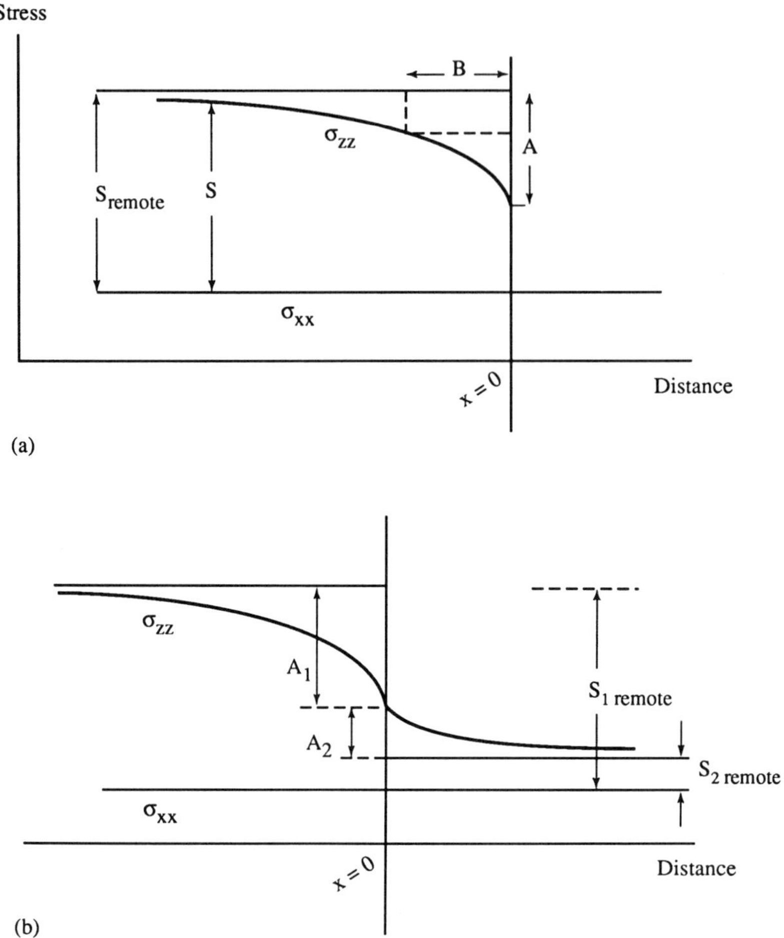

Figure 13.3 Exponential change of the compressive stress σ_{zz} near an interface.

In parallel with the above, $B_2 = (2N_2K_2)^{1/2}$ and from Figure 13.3b $A_1 + A_2 = S_{1\ remote} - S_{2\ remote}$, but we have not yet pinned down a definite magnitude for A_1 or A_2. The fact that enables us to determine these separately is that the flux to the interface from the left must equal the flux away from the interface toward the right. (If the interface were only weakly bonded, material could perhaps stack up *at* the interface as some kind of surface layer, and the flux-balance condition just noted would not need to be satisfied; but for the present, we assume perfect bonding and hence equal fluxes.)

The total separation $A_1 + A_2$ is fixed thus: from eqn. (13.4a) or from general mechanics,

$$S_{1\ \text{remote}} = 6N_1 e_0 \qquad \text{and} \qquad S_{2\ \text{remote}} = 6N_2 e_0$$

Hence

$$A_1 + A_2 \quad \text{or} \quad S_{1\ \text{remote}} - S_{2\ \text{remote}} = 6e_0(N_1 - N_2) \tag{13.6}$$

Also, the gradient $\partial\sigma_{zz}/\partial x$ or $\partial S/\partial x = (Ae^{x/B})/B$. At the interface this becomes A/B, or specifically A_1/B_1 on the left and A_2/B_2 on the right. Now, diffusivities are K_1 and K_2, so for equal fluxes we need

$$\frac{A_1 K_1}{B_1} = \frac{A_2 K_2}{B_2} \tag{13.7}$$

Equations (13.6) and (13.7) together give

$$A_1 = \frac{B_1 K_2}{B_1 K_2 + B_2 K_1} \, 6e_0(N_1 - N_2),$$

$$A_2 = \frac{B_2 K_1}{B_1 K_2 + B_2 K_1} \, 6e_0(N_1 - N_2) \tag{13.8}$$

A pair of materials with $B_1 = B_2$ is of particular interest; in such a pair

$$\frac{A_1}{A_2} = \frac{N_1}{N_2} = \frac{1/K_1}{1/K_2} \tag{13.9}$$

These are rather natural-seeming results; N_1 and $1/K_1$ can be thought of as measures of the resistance in material 1 to any kind of motion. If, in this sense, material 1 is more inert or resistant than material 2, it is reasonable to expect that the stress-drop across the neighborhood of the interface will be mainly in material 1, i.e., A_1 will be larger than A_2.

Orders of magnitude. It was remarked at the start of the chapter that, for purposes of mechanics, Figure 13.1b is often sufficient; the extra refinement in going to Figure 13.1c is often not needed. Then we naturally ask, when *do* we need to pay attention to diagram (c)? Turning to Figure 13.3a, how large might be the length B and the stress-difference A? A short answer is that in materials science, A can be hundreds of megapascals and B from 1 nm to 1 µm or more; in biological materials, A is smaller but B is still from 1 nm upward. These are *approximate* statements, perhaps no better than conjectures, because materials that come close to the simple behaviors the equations describe are hard to find: of materials whose behavior has been determined, most behave in more complicated ways.

The magnitude of B is indicated simply enough by eqn. (13.5), but the quantities N and K are not directly known. In the theory, N is a Newtonian coefficient linking stress difference to strain rate in conditions where their relation is *linear*—whereas experiments are mostly performed outside the linear range (at higher magnitudes of stress difference and strain rate); one has to extrapolate from the data in search of an estimate of what the behavior would reduce to if one could get down to the linear range. Estimates of K are uncertain for a different reason. The common way of estimating a self-diffusion coefficient is by adding a radioactive isotope; any step or anomaly in the distribution of the isotope will tend to smooth out with time, and the self-diffusivity is the coefficient that describes the rate at which the smoothing is seen to occur. But we wish to use a coefficient that describes *stress-driven* self-diffusion. The latter has never been directly measured (as far as I know); one coefficient is *in theory* related to the other by the factor $RT/$(molar volume), as in the Note, page xix, but as with the viscosity N, actual behavior might fail to be as simple as the equations imply.

With these cautions in mind, data are listed as follows (diffusivities are from Freer, 1981):

Glass at 1100 K. Assume that the glass deforms by movement of single oxygen atoms.

N	Viscosity, 10^5 Pa-sec
D^i	Isotope self-diffusion coefficient for oxygen atoms, 10^{-14} m²/sec
V	Volume of 1 kg-atom of oxygen, 14×10^{-3} m³/kg-atom
R	Gas constant, 8×10^3 J/kg-atom-K
B	$(2ND^iV/RT)^{1/2}$, 2.5 nm

Oxygen atoms in glass form tetrahedral clusters around silicon atoms. The edge of such a tetrahedron comprises two oxygen atoms touching each other. The overall length of such a pair is two diameters, or about 0.6 nm.

Compacted polycrystalline olivine at 1700 K. Assume that the polycrystalline aggregate deforms by movement of single oxygen atoms.

N	Viscosity, 10^{18} Pa-sec
D^i	Isotope self-diffusion coefficient for oxygen atoms, 10^{-18} m²/sec
V	Volume of 1 kg-atom of oxygen, 14×10^{-3} m³/kg-atom
R	Gas constant, 8×10^3 J/kg-atom-K
B	$(2ND^iV/RT)^{1/2}$, 45 μm

The deformation is believed to occur mainly by climb of dislocations, with oxygen atoms traveling along dislocation loops. Lengths of dislocation loops vary greatly but are of the order of magnitude of 1 μm.

Discussion

The situation just described deserves attention. Although it is very simple, it carries within itself half the content of this book (Chapter 14, where chemical inhomogeneity enters, carries the other half). To give a more realistic description of an actual interface between natural or man-made materials, one would have to introduce various complications, but no more novelties; the extra complications would be ones that people have already learned to deal with in other contexts. The simplifying assumptions made above lead to a problem that contains no mysteries, and yet it illustrates the fundamental thought from Chapter 11: *When a material deforms it behaves as if it contained 90°-arcs of finite length that serve as migration paths, and it is not reasonable to suppose that migrating material can behave differently on such a migration path according to whether the path is curved or straight.* That thought provides Figure 12.1e, eqn. (12.6), and the more general eqn. (12.7). Because of its value as a clean example, the interface problem will next be discussed from several points of view, but readers can jump from this point to Chapter 14 without losing the main thread.

Fletcher's problem. Fletcher's work on *slightly* porous, permeable granular aggregates was mentioned in Chapter 10, but now some more details are of interest. The situation he discussed is basically as in Figure 13.1a but on a scale of centimeters rather than nanometers. Each material is a homogeneous, coherent mass of granules with a porosity of one or two percent. In each, the dominant constituent is quartz grains, but because of differences in the other, less abundant constituents, aggregate 2 is more readily deformable than aggregate 1. The discussion is on a geological time scale; strain rates are of the order of 10^{-14} per second or 0.3 per million years; the rocks are assumed to behave mechanically as creeping fluids with viscosities of the order of 10^{20} Pa-sec.

The self-diffusion part of the behavior is taken to occur by means of a stationary intergranular watery film. This film allows quartz to dissolve at sites of high compression, travel through the film as a solute, and reprecipitate as quartz at sites of low compression. Interest attaches to (i) the mathematics and (ii) the comparison with natural examples of real rocks in analogous situations.

As regards the mathematics, Chapter 12 follows Fletcher's treatment closely. Fletcher derived exponential profiles; he was, I believe, the first person to do so by combining a change-of-shape term driven by the stress itself with a change-of-volume term driven by a spatial stress gradient [and fixed in magnitude by the second derivative, as in eqn. (12.6)]—the material that changes shape and the material that diffuses being essentially all the same material. In Fletcher's equations, the aggregates have the mechanical attributes of ideal continua, just as in Chapter 12.

Observations. Because of the scale of Fletcher's problem, it is possible to make observations comparable with the theory just by looking at a suitable pair of rock units with a naked eye; Figure 13.4 shows the kind of feature that can be observed:

1. Dissolution is at least partly localized. The dissolution sites shown in diagram (a) are concentrations of insoluble material; minor amounts of minerals other than quartz remain as residues when quartz dissolves.
2. Reprecipitated quartz is present in thin films distributed sporadically through both rock-types.

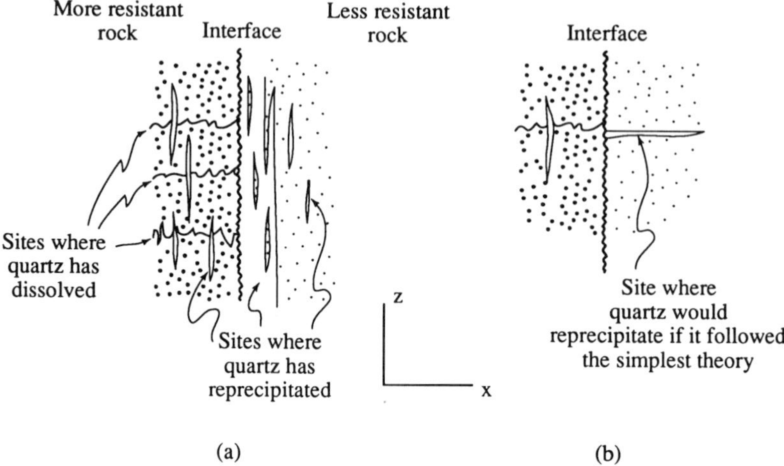

Figure 13.4 Features at an interface in natural rocks: (a) as actually observed; the region shown might be from 1 cm to 1 m across. (b) The conjectured form at a perfectly welded interface.

3. A substantial quantity of reprecipitated quartz, a layer several centimeters thick, is present at the boundary between the rock types.

These features partly accord with the theory and partly do not:

1. The films of residue and the sporadic films of reprecipitated quartz accord with the idea of shortening along z and elongation along x; at least part of the overall change of shape seems to occur by solution and reprecipitation.
2. The films of residue being parallel to each other accords with the idea that stress-driven loss of volume is not isotropic; the rock seems to show that volume-loss is a response specifically to σ_{zz} and is achieved by a shortening e_{zz}.
3. The thick accumulation of quartz at the interface is definitely not in accord with the theory. If the interface was as well bonded and coherent as is assumed, films of reprecipitated quartz would be sought on the downstream side of the interface as shown in Figure 13.4b. Such features sometimes appear at bent interfaces, but at planar interfaces they are very infrequent; a deposit as suggested in diagram (a) as far more commonly seen. As a second opportunity to practice with eqn. (12.6), behavior where the interface is *not* strongly bonded is discussed as follows.

Effect of a weak interface. The effect can be approximated by imagining a thin sheet of a third material occupying the interface. Let it have a low viscosity and high diffusivity compared with either material 1 or material 2. Because of its low viscosity, σ_{zz3} is always close to σ_{xx}. A profile in the

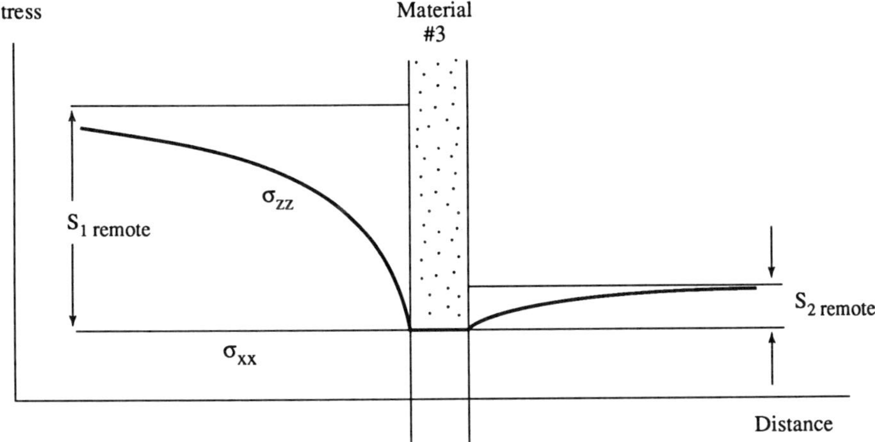

Figure 13.5 Profiles of the stress σ_{zz} near an interface that is occupied by a boundary layer of weak material.

manner of Figure 13.3b can then be drawn; see Figure 13.5. If the interface were indeed as weak as is assumed in this idealization, the diffusive flux in material 1 would be stronger, and the flux in material 2 would be completely reversed. For a picture of the geometry, refer again to Figure 13.2d. That figure shows the single interface migrating to the left (with respect to a coordinate frame that is pinned to the materials at remote points) as material diffuses out of phase 1. In the weak-interface condition, the left-hand limit of material 2 behaves independently: it retreats to the right at its own pace rather than following material 1. Material 3 accumulates, and its width increases, at a rate which is the sum of effects on the right-hand and left-hand sides.

We should perhaps note that the behavior at remote points is exactly the same whether the interface is ideally bonded or ideally weak. The interface could in fact oscillate between the two extremes without behavior at remote points being affected in any way. Again, rocks provide what seem to be large-scale natural examples; some veins are known that seem to have oscillated between being sealed up and being freely open ten or more times in their course of development. But in these real examples, conditions are probably far from those assumed by Fletcher, as regards the intergranular fluid film standing still while just the solute moves from place to place.

Plane strain. For an instructive variation on the problem discussed above, consider replacing the condition of cylindrical symmetry with the condition of plane strain; that is, assume $e_{yy} = 0$ instead of assuming $e_{yy} = e_{zz}$.

In the elementary viscous mechanics of nondiffusing materials, the condition $e_{yy} = 0$ goes with the stress condition $\sigma_{yy} = (\sigma_{zz} + \sigma_{xx})/2$. For example,

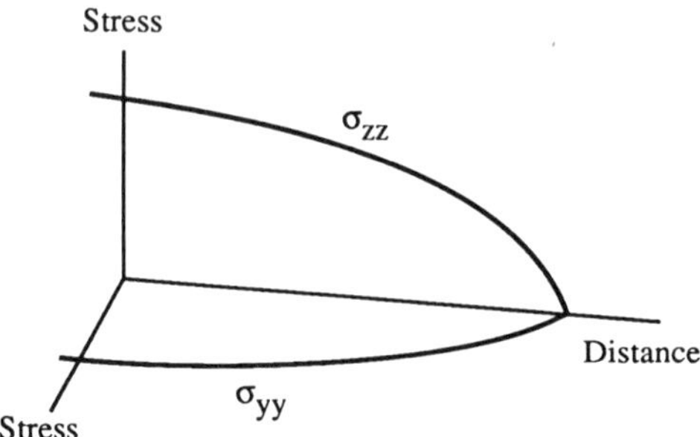

Figure 13.6 Profiles of σ_{zz} and σ_{yy} that satisfy the condition $\sigma_{yy} = (\sigma_{zz} + \sigma_{xx})/2$, where σ_{xx} is uniform.

if σ_{xx} were constant and σ_{zz} diminished along an exponential curve as in Figure 13.5, left-hand side, then to maintain $e_{yy} = 0$, σ_{yy} would need to diminish along an exponential curve with just half the amplitude, as in Figure 13.6. But if the material is diffusive, the simple proportionality between σ_{yy} and σ_{zz} disappears. The algebra is given in Appendix 13A but is of less concern than the conceptualization: if σ_{yy} varies along x, material will diffuse to the right for this reason and there will be shortening along y proportional to $\partial^2\sigma_{yy}/\partial x^2$; then to maintain the condition $e_{yy}^{\text{total}} = 0$, we need a change of shape in the yz plane that did not appear in the cylindrically symmetrical problem—a shortening down z and elongation along y. To permit this change of shape, σ_{yy} needs to drop *below* the value $(\sigma_{zz} + \sigma_{xx})/2$, by different amounts at different points along the x-direction. Hence the simple proportionality between the profiles for σ_{zz} and σ_{yy} disappears; both have to take on shapes that are no longer simple exponential curves.

Without showing the algebra here, we note that whereas eqn. (12.6a) suffices for the cylindrically symmetrical problem, for the plane-strain problem we use three second derivatives from the full set of five:

$$e_{zz} = \frac{1}{12N}\frac{\partial^2\sigma_{zz}}{\partial\alpha_1^2} + \frac{1}{12N}\frac{\partial^2\sigma_{zz}}{\partial\alpha_2^2} + \frac{K}{3}\frac{\partial^2\sigma_{zz}}{\partial x^2} \qquad (13.10)$$

But the nature of the problem still leaves variations along directions y and z at zero.

The factor f. An unsatisfactory feature in eqn. (12.7) is the factor f. In some circumstances it is of little significance, but clarifying its role will be a continuing task through the remainder of the book.

The reason that f can sometimes be safely ignored is illustrated by the two interface problems just discussed. Inside one homogeneous phase, gradients in stress magnitude from point to point can develop only to a limited extent; introducing a phase boundary immediately changes the situation and opens up the possibility of almost infinite gradients in σ_{zz} and

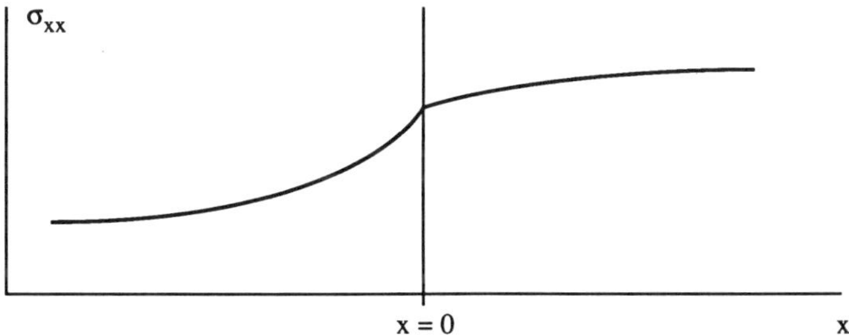

Figure 13.7 The manner in which σ_{xx} varies along x. Such a variation must be present, but is small enough to be neglected in a first analysis, in comparison with variations in σ_{yy} and σ_{zz}.

σ_{yy} (for an interface normal to x) but *not* in σ_{xx}—even at the interface, there is little reason for σ_{xx} to change. Thus $\partial^2\sigma_{zz}/\partial x^2$ can be large, but $\partial^2\sigma_{xx}/\partial x^2$ in general cannot; or using a general direction $\mathbf{n}$, as in eqn. (12.7), one can always make $\partial^2\sigma_{nn}/\partial x_l^2$ large by introducing an interface normal to l, but one cannot make $\partial^2\sigma_{nn}/\partial x_n^2$ large by introducing an interface normal to $\mathbf{n}$.*

An assumption made earlier deserves to be recalled at this point. In deriving the form of σ_{zz} in Figure 13.3, etc., it was assumed that σ_{xx} is the same at all points throughout the sample. This can be true only to the extent that the diffusive movements occur without dragging on the host material. Referring to Figure 13.2e, if the diffusive movements shown there in fact exert some drag, the profile of σ_{xx} must be of the form shown in Figure 13.7. That is to say, $\partial^2\sigma_{xx}/\partial x^2$ certainly becomes nonzero (and is infinite at the interface, at the discontinuity in slope). The idea that $\partial^2\sigma_{xx}/\partial x^2$, while not quite zero, is likely to be much smaller than $\partial^2\sigma_{zz}/\partial x^2$ is supported by an argument based on Chapter 11 as follows.

Consider for simplicity a plane strain $e_{zz} = e$, $e_{xx} = -e$ driven by a stress difference $\sigma_{zz} - \sigma_{xx} = S$. The rate at which energy is transformed is eS per unit volume. If the change of shape occurred by the imaginary process in Figure 11.5, the volume of migrating material would be e m^3 per cubic meter per second. Imagine the motion to be in discrete wafers, each having volume w; then the number that move equals e/w per m^3-sec and the energy transformed by each wafer's migration is wS. The path traveled has length L_0, so an imaginary force is associated with migration, of magnitude wS/L_0.

At this point, we transfer to Figure 11.6. Because of the similarity emphasized in Chapter 11, the same force is associated with movement of a wafer along x in the presence of a gradient $\partial\sigma_{zz}/\partial x$. The gradient in Figure 11.6 runs from a minimum of zero to a maximum of $(\pi/2)S/L_0$ with an average magnitude S/L_0 so that here the imaginary force has a magnitude of the order of $w\,\partial\sigma_{zz}/\partial x$. Let there be n such wafers present per m^3 of host: then the total force per m^3 of host that arises from diffusion is $nw\,\partial\sigma_{zz}/\partial x$

---

* An exception to this statement appears in the following paragraph.

and this equals the gradient of interest, $\partial\sigma_{xx}/\partial x$. The factor nw is the total volume of material that at any instant is engaged in diffusion, per m^3 of host material.

The result just gained seems reasonable both for stiff condensed continua such as glass and for gases. In a deforming gas, every molecule is engaged in diffusion all the time as well as being engaged in deformation; $nw = 1$ and we expect only small deviations from the state $\sigma_{zz} = \sigma_{xx}$, $\partial\sigma_{zz}/\partial x = \partial\sigma_{xx}/\partial x$. On the other hand, as discussed on page 57, for a glass, at any moment most of the material is behaving as a coherent continuum and only a small volume-fraction is engaged in self-diffusion, and correspondingly $\partial\sigma_{xx}/\partial x$ is very small compared with $\partial\sigma_{zz}/\partial x$.

All these remarks are of course concerned with interfaces where no chemical contrast is present. The effect of diffusion on the profile of σ_{xx} across an interface with contrast in chemistry is discussed in Chapter 16.

Summary

If two polymorphs are strongly bonded at a planar interface and subjected to a constriction stress field about the interface's normal as the constriction axis, for the strain to be uniform the compressive stress profile must be exponential. Within a few nanometers of the interface, stress gradients and diffusion of material become noticeable. Equation (12.6) applies and Figure 13.3 illustrates the results.

For plane strain, if the two polymorphs deformed without diffusion, σ_{yy} would need to equal $(\sigma_{zz} + \sigma_{xx})/2$; but if self-diffusion occurs, this stress condition no longer leads to plane strain. Where cylindrical constriction requires a simple exponential profile of compressive stress, plane-strain deformation at an interface requires a more complicated curve.

The analyses emphasize the idea that when a volume-element changes volume by diffusive mass transfer at constant density, the change is not isotropic; the strain components e_{xx}, e_{yy}, and e_{zz} are linked by three separate equations to the components and gradients of the driving stress field.

The fifth right-hand term in eqn. (12.7) contains a factor that is not fully defined, but for applications of the type illustrated no difficulty arises because the entire fifth term is very small in comparison with the other four.

APPENDIX 13A: PLANE STRAIN AT AN INTERFACE

The strain conditions to be imposed are $e_{zz} = e_0$ and $e_{yy} = 0$ at all points. Otherwise conditions are as in the constriction problem; in particular we continue to assume that no quantities vary along y or z. We apply eqn. (12.7) first with $\mathbf{n}$ along z and then with $\mathbf{n}$ along y, thus:

$$e_{zz} = e_0 = \frac{1}{12N}\left[\frac{\partial^2 \sigma_{zz}}{\partial \alpha_1^2} + \frac{\partial^2 \sigma_{zz}}{\partial \alpha_2^2}\right] + \frac{K}{3}\frac{\partial^2 \sigma_{zz}}{\partial x^2} \qquad (13A.1a)$$

$$e_{yy} = 0 = \frac{1}{12N}\left[\frac{\partial^2 \sigma_{yy}}{\partial \alpha_1^2} + \frac{\partial^2 \sigma_{yy}}{\partial \alpha_2^2}\right] + \frac{K}{3}\frac{\partial^2 \sigma_{yy}}{\partial x^2} \qquad (13A.1b)$$

As before let α_1 and α_2 lie in the xz and yz planes, respectively; then

$$e_0 = -\frac{1}{6N}\left[(\sigma_{zz} - \sigma_{xx}) + (\sigma_{zz} - \sigma_{yy})\right] + \frac{K}{3}\frac{\partial^2 \sigma_{zz}}{\partial x^2} \qquad (13A.2a)$$

$$0 = -\frac{1}{6N}\left[(\sigma_{yy} - \sigma_{xx}) - (\sigma_{zz} - \sigma_{yy})\right] + \frac{K}{3}\frac{\partial^2 \sigma_{yy}}{\partial x^2} \qquad (13A.2b)$$

We introduce stress differences R and T thus:

$$R = \frac{(\sigma_{zz} + \sigma_{yy})}{2} - \sigma_{xx}, \qquad T = \frac{\sigma_{zz} - \sigma_{yy}}{2}$$

Then half the sum of (13A.2a) and (13A.2b) gives

$$\frac{e_0}{2} = -\frac{R}{6N} + \frac{K}{3}\frac{\partial^2 R}{\partial x^2}, \qquad \frac{\partial^2 R}{\partial x^2} = \frac{1}{2NK}(R_{\text{remote}} - R) \qquad (13A.3)$$

and half the difference of (13A.2a) and (13A.2b) gives

$$\frac{e_0}{2} = -\frac{T}{2N} + \frac{K}{3}\frac{\partial^2 T}{\partial x^2}, \qquad \frac{\partial^2 T}{\partial x^2} = \frac{3}{2NK}(T_{\text{remote}} - T) \qquad (13A.4)$$

The solutions of (13A.3) and (13A.4) are similar to each other and to the solution of (13.4), but the value of the parameter B that results from (13A.4) is new and differs from that for (13A.3). In other words, T diminishes with distance from the interface on a different length scale from R.

To work through the details, let $R = A_R e^{x/B_R}$ and $T = A_T e^{x/B_T}$. Across the interface, we impose two flux conditions separately: first, the fluxes on either side driven by $\partial\sigma_{zz}/\partial x$ must balance, and also the fluxes on either side driven by $\partial\sigma_{yy}/\partial x$ must balance. These give the result that $A_{R1} = 3A_{T1}$ and $A_{R2} = 3A_{T2}$.

In the cylindrically symmetrical situation, the conclusion was that at any point along the $-x$ direction,

$$\sigma_{yy} = \sigma_{zz} = \sigma_{zz}^{\text{remote}} - Ae^{x/B}$$

but for plane strain the results are

$$\sigma_{yy} = \sigma_{yy}^{\text{remote}} - A_{T1}(3e^{x/B_{R1}} - e^{x/B_{T1}})$$

$$\sigma_{zz} = \sigma_{zz}^{\text{remote}} - A_{T1}(3e^{x/B_{R1}} + e^{x/B_{T1}})$$

If B_{T1} were equal to B_{R1}, these would satisfy the condition $\sigma_{yy} = (\sigma_{zz} + \sigma_{xx})/2$, but because B_R is larger than B_T by the factor $3^{1/2}$, the simple proportionality of σ_{yy} is lost.

14

ONE PHASE AND TWO COMPONENTS

The purpose of this chapter is to bring chemical variation into the examples discussed. Equation (12.7) shows a material's change of dimension as the sum of two effects, change of shape and diffusive mass transfer, but it describes only changes driven by unequal stress components. Equation (12.8) *suggests* that the same form continues to be applicable when potential gradients exist for reasons that are not stress-related, but the suggestion is not followed up and Chapter 13 is entirely concerned with effects of stress. We now go back to the harmonic variation in Figure 11.6 and ask: supposing that both the radial compression *and the composition* fluctuate along the x-direction, what material movements will result? An answer to this seemingly simple and fundamental question was produced rather recently, by G. B. Stephenson, whose work provides the foundation for this chapter.

Initial conditions and assumptions

At once we have to give up the idea of a continuum and turn to a material containing countable numbers of atoms. Let the atoms be of just two species A and B; let the number of atoms of A in 1 m^3 be ρ_A and the volume occupied by each one be V_A, and similarly for B: then

$$\rho_A V_A + \rho_B V_B = 1 \qquad (14.1)$$

The relative amounts of A and B can be expressed in various ways: their number-ratio is just ρ_A/ρ_B, their volume-ratio is $\rho_A V_A/\rho_B V_B$ and their mass-ratio, which we do not need, would be different again. Also the concentration of A can be taken as $\rho_A/(\rho_A + \rho_B)$ and written C_A: of any large number of atoms, a fraction C_A of that number would be expected to be atoms of A. Going back to Chapter 3, especially eqns. (3.4), we may need to distinguish the *activity* of species A from its concentration, but before embarking on those details, we can already consider the general nature of the problem and its solution.

Two conditions or experiments can be readily imagined.

(1) Imagine a long wormlike sample in air. Let the ratio C_A vary harmonically along the length and let atoms of A be larger than atoms of B (Figure 14.1). If no other influences bear on the sample, interdiffusive movements of atoms will occur; portions formerly poor in A will gain A and lose B, portions formerly poor in B will do the opposite, and in general the

129

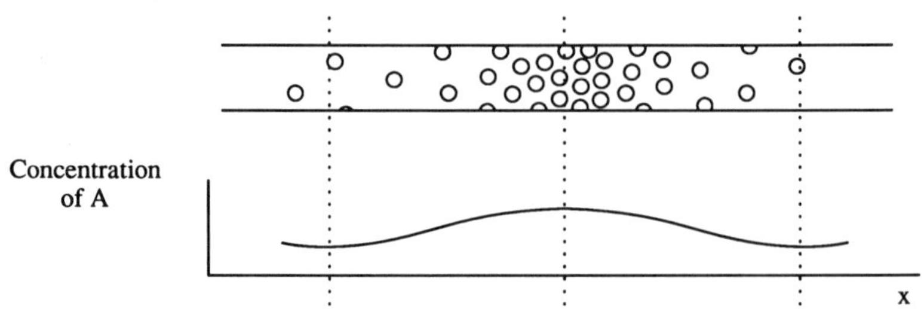

Figure 14.1 A cylindrical sample with harmonic variation in content of component A.

radius of the worm will change, swelling at some points and narrowing at others. The changes in radius depend on the mobilities of the two species, as can be seen by considering two extremes: if A is far more mobile than B, portions originally poor in A will be swelled by an influx of A, whereas if B is far more mobile than A, portions originally poor in A will shrink through an efflux of B.

Will the state of stress in the worm be affected? The radial compression will continue to be atmospheric pressure, and we can imagine the worm to be floating on an inviscid fluid support (such as mercury). Then if the axial compression is uniform at the start, *there is no reason for it to become nonuniform.* For example, consider the second instance just imagined: B is the more mobile species so that a region where the concentration of A was initially a maximum will be invaded by B from both sides. Will B drag on A as it arrives, thus increasing the axial compression on the central element? No, attention has to be paid to the fact that the concentration gradient makes B move *with respect to A* but not with respect to any external reference frame, so no build-up of compression occurs.

This fundamental point deserves review. One approach is to give attention to momentum. Suppose that at the initial moment, the worm overall has no momentum along the direction of its length, and no subdivision has any momentum either. Atoms of A and B can *exchange* momentum, but however much they do this, an element of worm that had no momentum at the start cannot pick up momentum from anywhere.

Specifically, consider a volume element that is a thin slice with parallel faces normal to the worm's centerline: such an element's momentum could be changed by a lengthwise drag on its cylindrical external surface, or by filling it with magnetic particles and imposing an *external* magnetic field— but internal exchanges of momentum among As and Bs can only leave the net momentum at zero.

(2) Next, imagine the worm-like sample to be fitted into a perfectly lubricated rigid cylindrical tube. Again we suppose that at some initial moment the composition varies harmonically, so that again a region initially low in B is invaded by B from both sides. In the new circumstances, the

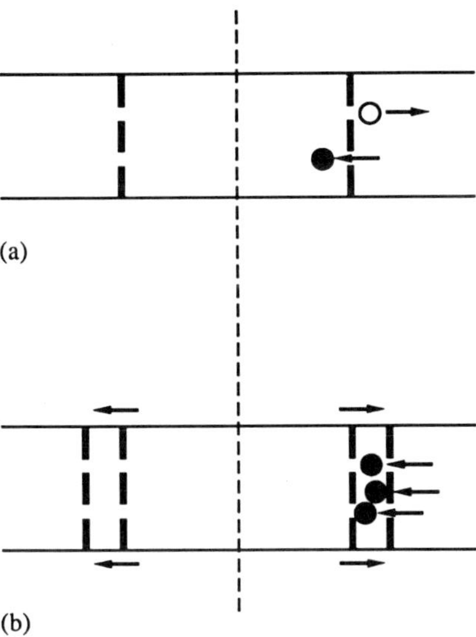

Figure 14.2 The movement of atoms of B into an element of space. (a) If the boundaries of the element stay fixed, an equal volume of material A must move out. (b) If the element boundaries move, the content of A can remain constant while the content of B increases.

central element cannot expand radially to admit the new atoms of B: the increase in concentration of B can occur only by an equal *volume* of A moving out, if we give attention to an element whose boundaries stay fixed while A and B travel across them (Figure 14.2a). But an alternative useful view is as follows: let the central element be defined by the atoms of A it originally contained; then as atoms of B enter, the original atoms of A have to disperse over a greater length of tube (Figure 14.2b). In the region considered, a swarm of B atoms shortens and occupies a greater fraction of the width of the tube, while a swarm of A atoms gets longer and occupies a smaller fraction of the width of the tube. We are currently assuming that A is the less mobile of the two components—in experiment (1), the swarm of A elongated and thinned a *little* by its own mobility but not by as much as B thickened; then the extra elongation of A that occurs in experiment (2) is accompanied by a radial compressive stress at the frictionless walls. Whether one says the stress causes the behavior of A or the behavior of A causes the stress is a matter of choice; or perhaps one should say the behavior of B causes both the stress and the behavior of A. In any case, the result is a stress profile of the general type shown in Figure 14.3.

To work out completely the relation of the stress profile to the composition profile, we would have to find its form as well as its amplitude; but to see the nature of the problem, so as to begin seeking a solution, we can focus on the amplitude, assuming the profile to be something like a sine wave. The

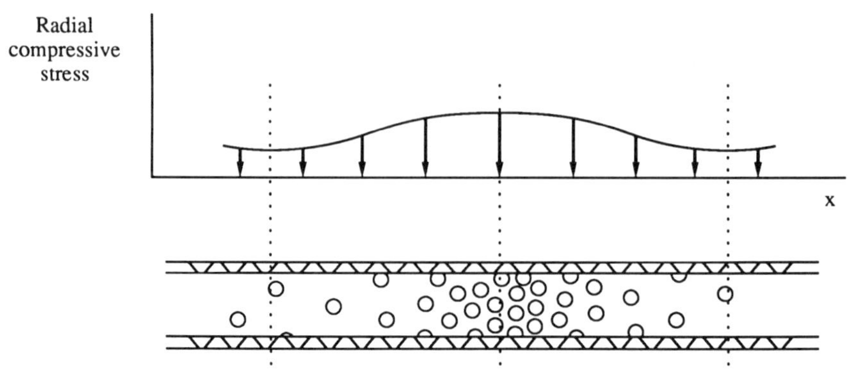

Figure 14.3 Variation in radial compressive stress accompanying a variation in composition. The material is imagined as confined in a frictionless cylindrical tube of fixed width.

point to note is that a compression profile of the type shown not only hastens the outward movement of A, but also opposes the inward movement of B. The condition we seek is one where mobile B, driven by a concentration gradient but *hindered* by a compression gradient, moves into an element of fixed size at just the same rate (in m^3/sec) that less mobile A moves out, driven by a concentration gradient *plus* a compression gradient; a fluctuation in compression must develop that just equalizes the two volume-fluxes.

Before embarking on the solution, there are two points to note: (1) The condition that σ_{xx} is uniform along the length of the sample continues to hold (taking the x-direction along the sample's length). The confining tube is taken as ideally frictionless and capable of exerting only radial compressive stress; an element continues then to have no means of gaining any momentum parallel to x. (2) There is a whole series of pairs of composition profile plus accompanying stress profile that satisfy the condition we seek. In particular, suppose we start with a composition profile of amplitude C and find that its accompanying stress profile has amplitude S; then a composition profile with a slightly smaller amplitude C' will presumably be accompanied by a stress profile of slightly smaller amplitude S'. We do not seek a stress fluctuation sufficiently strong to bring interdiffusion to a standstill; we only seek to adjust the interdiffusion so as to fit the fixed-radius rigid-confinement condition. Interdiffusion then does proceed: the profile C gradually flattens with time to become C', and simultaneously the stress profile flattens from S to S'. The system moves toward uniformity of both composition and stress, down a line or path of compatible C,S pairs.

(And Stephenson's result is that if the C profile is a sine wave of small amplitude, the S profile is indeed a sine wave as well. The next section gives the algebra to establish this result.)

Solution

The treatment starts by assuming stresses as in Figure 11.6 and composition as in Figure 14.1, thus:

$$\sigma_{zz} = \sigma_{xx} + s \sin \theta \qquad (14.2)$$

where $\theta = jx$ and the wavelength is $2\pi/j$,

$$\frac{\rho_A}{\rho_A + \rho_B} = C^A = C_0^A + c \sin \theta \qquad (14.3a)$$

and

$$C^B = (1 - C_0^A) - c \sin \theta \qquad (14.3b)$$

That is to say, we begin by assuming that σ_{zz} is harmonic, with just the amplitude s as the only unknown; we find that a solution is possible along these lines that makes a starting point for further consideration of the problem.

As in Chapter 13, there is cylindrical symmetry and no variation of any quantity along y or z, so that eqn. (12.8) reduces to two terms thus:

$$e_{zz} = D\left(\frac{\partial^2}{\partial(L\alpha)^2} + \frac{\partial^2}{\partial x^2}\right)\mu_z \qquad (14.4)$$

Here the material's characteristic length $2(NK)^{1/2}$ is written as L rather than R to avoid confusion with the gas constant. We write the potential μ_z (in joules/mole) as $\mu_z^0 + V\sigma_{zz} + RT\ln C$, and so generate four terms on the right-hand side:

$$e_{zz} = D\left(\frac{V}{L^2}\frac{\partial^2\sigma_{zz}}{\partial\alpha^2} + \frac{RT}{L^2}\frac{\partial^2(\ln C)}{\partial\alpha^2} + \frac{V}{L^2}\frac{\partial^2\sigma_{zz}}{\partial x^2} + RT\frac{\partial^2(\ln C)}{\partial x^2}\right) \qquad (14.5)$$

We take the concentration of a species within a small volume to be an isotropic quantity so that the second term is thrown out; also, as before, $\partial^2\sigma_{zz}/\partial\alpha^2 = -2(\sigma_{zz} - \sigma_{xx})$ (from page 101). Hence

$$\text{shortening} \qquad e_{zz} = D\left(\frac{V}{L^2}2s \sin \theta + Vj^2s \sin \theta \pm RTj^2 \frac{c}{C_0} \sin \theta\right) \qquad (14.6)$$

[The last term in (14.5) gives just the last term in (14.6) as long as c is small in comparison with C in (14.3).]

The last term in (14.6) has two signs because the next step is to recall that we have two separate components present. As a tentative step, we write (14.6) separately for each component, and then consider what the equations might represent in physical terms. Thus:

$$e_{zz}^{\mathrm{A}} = \frac{s \sin \theta}{6N^{\mathrm{A}}} + D^{\mathrm{A}} V^{\mathrm{A}} j^2 s \sin \theta - RT \frac{D^{\mathrm{A}}}{C^{\mathrm{A}}} j^2 c \sin \theta \tag{14.7a}$$

$$e_{zz}^{\mathrm{B}} = \frac{s \sin \theta}{6N^{\mathrm{B}}} + D^{\mathrm{B}} V^{\mathrm{B}} j^2 s \sin \theta + RT \frac{D^{\mathrm{B}}}{C^{\mathrm{B}}} j^2 c \sin \theta \tag{14.7b}$$

Interpretation. The left-hand sides are fairly straightforward. The volume-fractions of the two components are $\rho^{\mathrm{A}} V^{\mathrm{A}}$ and $\rho^{\mathrm{B}} V^{\mathrm{B}}$. We abbreviate these as a and b (with $a + b = 1$); then on any surface normal to x, the *areas* of the components will be in the ratio $a{:}b$, and any *diameter* will be composed of length-segments with overall ratio $a{:}b$. If the fraction a strains by e^{A} and the fraction b strains by e^{B}, the overall strain is $ae^{\mathrm{A}} + be^{\mathrm{B}}$, and the essence of the problem is to keep this overall effect at zero.

When we multiply the right-hand sides by a and b, respectively, and add, again the total must be zero, and the six terms form three pairs thus:

$$\frac{s \sin \theta}{6} \left(\frac{a}{N^{\mathrm{A}}} + \frac{b}{N^{\mathrm{B}}} \right)$$

$$\frac{j^2 s \sin \theta}{3} (aK^{\mathrm{A}} + bK^{\mathrm{B}}) \qquad (\text{using } DV = K/3, \text{ p. } 103)$$

$$RTj^2 c \sin \theta \left(\frac{bD^{\mathrm{B}}}{C_0^{\mathrm{B}}} - \frac{aD^{\mathrm{A}}}{C_0^{\mathrm{A}}} \right) \qquad \text{or} \qquad \frac{j^2 c \sin \theta}{3} RT(\rho^{\mathrm{A}} + \rho^{\mathrm{B}})(K^{\mathrm{B}} - K^{\mathrm{A}})$$

Each pair represents a tendency in the sample to change radius. The third of the pairs is the composition-driven interdiffusion effect, where the two species travel in opposite directions, and the effect arises from their difference in mobility, $(K^{\mathrm{B}} - K^{\mathrm{A}})$. If they were of equal mobility they would simply exchange places volume-for-volume in a compatible way and no stress anomalies would develop, i.e., s would be zero. As already discussed, when the mobilities are not equal, the sample tends to swell at some points and get thinner at other points, and it is the business of the stress terms to just exactly nullify this effect.

Of the stress terms—the first two pairs above—the first is for change of shape at constant volume; "to make room for the incursion of mobile B, the material already present shortens radially and elongates axially." The anthropomorphic phrasing is inaccurate but convenient. The magnitude of the effect is controlled by the materials' two viscosities jointly, as shown by the sum of two terms involving N.

Similarly, the second stress term, or pair of terms, is for stress-driven diffusion; in addition to the change of shape, some material diffuses along x away from the high-stress sites where incursion of B is a maximum. Again the two materials work jointly; the effect depends on a weighted sum of their mobilities K. A difference is that this term, quite properly, increases if j

increases; that is to say, if the wavelength diminishes and the sink-regions come closer to the source-regions, all stress-driven diffusion fluxes increase; only the change-of-shape effect is independent of j.

In the condition for the sum to be zero, the common factor $(\sin \theta)/6$ plays no part; we need

$$s\left[\left(\frac{a}{N^A} + \frac{b}{N^B}\right) + 2j^2(aK^A + bK^B)\right] = c \cdot 2j^2 RT(\rho^A + \rho^B)(K^B - K^A) \quad (14.8)$$

At least, the condition that needs to be satisfied takes this general form. We can abbreviate it as

$$s\left[\frac{1}{N^*} + 2j^2 K^*\right] = c \cdot 2j^2 E^* \quad (14.9a)$$

or

$$s = \left(\frac{2j^2 E^*}{(1/N^*) + 2j^2 K^*}\right) c \quad (14.9b)$$

$$\frac{1}{N^*} = \frac{a}{N^A} + \frac{b}{N^B}, \qquad K^* = aK^A + bK^B, \qquad E^* = RT(\rho^A + \rho^B)(K^A - K^B)$$

If a small harmonic variation in a material's composition exists along some direction, this equation shows how large a harmonic variation in transverse compressive stress magnitude we can expect to exist at the same time.

Equation (14.9) is no more than a preliminary sketch at this stage. It contains two uncertainties: firstly it is not certain that to form, for example, K^* the best combination to use is $aK^A + bK^B$. Secondly, and more fundamentally, the terms containing RT in eqns. (14.5) onward presuppose two populations of *atoms* that interdiffuse by jumping past each other. Then we have to ask how valid is it to write e^A, the "strain rate of component A" as if component A formed a continuum, as in continuum mechanics. Chapter 15 sheds some light on this question, but does not resolve it completely. The whole remainder of the book is on an exploratory or tentative basis.

Despite its being in preliminary form, eqn. (14.9) is essential for succeeding chapters. It embodies two important ideas. First it describes a condition where the stress fluctuation s and composition fluctuation c are proportional to each other; it is this proportion that remains constant as the two fluctuations decay toward zero together. Second, eqn. (14.9a) clearly displays the three effects that interact—change of *shape* of the ensemble at a point, migration from high stress to low stress, and migration driven by concentration. The details that occupy the remainder of this chapter should not distract the reader; a mental image of the three processes running in rather concrete billiard-ball style is important at this stage.

Discussion

The first point to be considered is the profile shape of the transverse stress σ_{zz}. In an experiment, one might imagine controlling the profile shape of the concentration C, and *discovering* how σ_{zz} varied. Here we *assumed* that σ_{zz} is harmonic and derived eqn. (14.9) just to fix its amplitude. Looking at eqn. (14.7), one can see that *if* C is harmonic, then a similar harmonic variation in σ_{zz} is certainly one possible solution. To go further and show that this is the only possible solution would be an extra step. For the present, we simply pass over the possibility that other solutions might exist, and explore the significance of the solution we have in hand.

The way in which the properties E^*, K^*, and N^* combine to fix s becomes clearer if we use the idea of the material's characteristic length L or $2(NK)^{1/2}$. Whereas N and K are measures of activity or mobility, L is a more geometrical quantity; as discussed in Appendix 12A, it is related to the actual path an atom follows when it moves among other atoms. Thus, N and K are *strongly* sensitive to temperature, for example, whereas L is rather insensitive to temperature. We apply this thought to the term $2j^2(aK^{\mathrm{A}} + bK^{\mathrm{B}})$ on the left of eqn. (14.8): it can be rewritten

$$\frac{j^2}{2}\left(\frac{aL_{\mathrm{A}}^2}{N_{\mathrm{A}}} + \frac{bL_{\mathrm{B}}^2}{N_{\mathrm{B}}}\right)$$

We next propose that if A and B are similar enough in size and behavior to form the mixture we have in view, the two lengths L_{A} and L_{B} cannot be *greatly* different, and substitute as an approximation the term

$$\frac{j^2}{2}\,L_{\mathrm{m}}^2\left(\frac{a}{N_{\mathrm{A}}} + \frac{b}{N_{\mathrm{B}}}\right) \qquad \text{with } L_{\mathrm{m}} \cong L_{\mathrm{A}} \cong L_{\mathrm{B}}$$

The left-hand side of eqn. (14.8) becomes more unified, and eqn. (14.9b) becomes

$$s = \left(\frac{2j^2}{1 + j^2 L_{\mathrm{m}}^2/2}\right)E^*N^*\cdot c \tag{14.9c}$$

$$\frac{2j^2}{1 + j^2 L_{\mathrm{m}}^2/2} = \frac{4}{(2/j^2) + L_{\mathrm{m}}^2} = \frac{4}{(\lambda^2/2\pi^2) + L_{\mathrm{m}}^2}$$

For a prescribed magnitude of c, the magnitude of s is thus proportional to E^*, the difference in the components' mobilities, to N^*, their resistance to motion, and to the geometrical factor. For most values of λ, L_{m}^2 is negligible and s becomes proportional to $1/\lambda^2$; the larger λ, the less steep the concentration gradients that tend to drive motion, and hence the smaller the stress fluctuations that are produced. But when λ is of the order of $4L_{\mathrm{m}}$, the two parts of the denominator are equal, and if λ diminishes below this value, the factor as a whole does not increase so strongly.

In terms of physics, we have reached the stage where diffusion of A and B together away from high-stress sites is as important a relief mechanism as change of shape. At this stage, shortening the source-to-sink distance does not affect s strongly, because the relief mechanism is enhanced at the same time that the driving gradients are enhanced. [Going back to (14.9c), once $j^2 L_m^2/2$ is the dominant part of the denominator, increasing j is ineffective— the numerator and the denominator increase in the same proportion.]

The form of N^*. A further point of interest in eqn. (14.9) is the material property N^*, which measures the resistance of the mixture of A and B to change of shape or volume. It will be recalled that in forming eqns. (14.7), the components A and B were treated as geometrically independent: it was assumed that component A responds to s according to its viscosity N^A, that component B responds separately according to its viscosity N^B, and that the total effect is a weighted sum or average of these two separate strain responses. We should consider whether this is the best of possible approaches, as follows.

Suppose two fluids interpenetrate in some intricate manner, as suggested by Figure 14.4a. We might think of an overall stress difference s and an overall strain rate e (and for the purposes of this section, two dimensions suffice). Different behaviors emerge according to how these overall values partition themselves in the two materials.

Simple extreme cases are as follows. (1) Let the stress be imposed uniformly throughout both materials, and let their different viscosities give different local strain rates, as in Figure 14.4b. An average strain rate for the sample can be got by a geometrical procedure (approximating the irregular

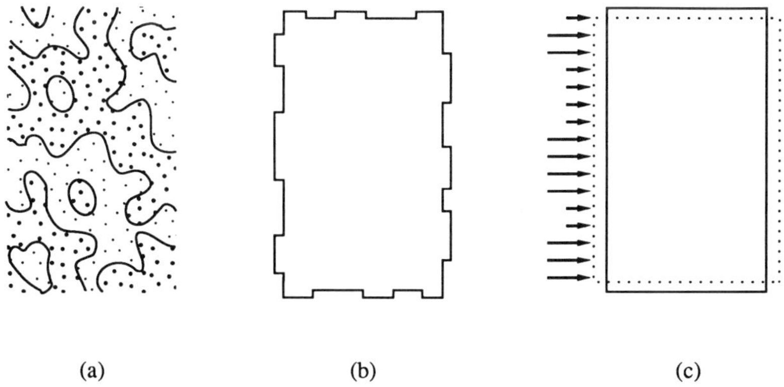

(a)   (b)   (c)

Figure 14.4 Interpenetration of two continuous phases with different viscosities. (a) A conceptual picture of an extended interpenetration; the same type of geometry exists in three dimensions. (b) If a stress field is imposed uniformly over both phases, the velocity fields in the two phases will be different; initially planar surfaces become nonplanar. (c) If a strain field is imposed uniformly over both phases, the stress field is necessarily different in the two phases.

boundaries by straight lines, or by an equivalent algorithm expressed in more abstract terms), and it is $ae^A + be^B$ if each component by itself deforms homogeneously (with strain rate e^A at every point within component A). In this case, the effective viscosity overall is $s/4(ae^A + be^B)$; but separately $e^A = s/4N^A$ so the effective viscosity overall is $1/(a/N^A + b/N^B)$ as already adumbrated.

(2) Let the *strain* be imposed uniformly throughout both materials, as in Figure 14.4c; then if this could be done by having two separate stress fields, one homogeneous throughout material A and the other homogeneous throughout material B, we should need $s^A = 4eN^A$, $s^B = 4eN^B$, and $as^A + bs^B = s$. In this case the overall viscosity $s/4e$ is $aN^A + bN^B$.

Neither of these extreme cases really works: the first involves geometrical mismatch at the internal boundaries, with, for example, B moving into space that is already occupied by A, and the second involves boundaries between materials A and B where the stress on one side is not equal to the stress on the other side. However, the general result is acceptable—that the governing relation is

$$N^* = aN^A + bN^B \qquad \text{or} \qquad \frac{1}{N^*} = \frac{a}{N^A} + \frac{b}{N^B}$$

or somewhere between these two extremes, according to whether the ensemble behaves more like Figure 14.4b or 14.4c. The remarks made on page 91 in connection with Figure 11.4 in fact emphasized that *surfaces* in the continuum remain continuous; to the extent that this approach is correct, we should favor diagram (c) over diagram (b), and favor the left-hand form of N^* rather than the right-hand form. But obviously Figure 14.4a is a poor representation of a mixture of atoms, and the proper treatment of N^* remains not clear-cut.

Stress-driven changes. Up to this point, the entire chapter has been based on the idea of the chemical fluctuation being the "cause" and the stress fluctuation a consequence. But eqn. (14.9) applies equally well if the stress fluctuation s is independently established; then the question that is answered by eqn. (14.9) is: Given the stress fluctuation (which in the normal course would produce *various* transverse shortening rates at different points along the sample), what chemical fluctuation needs to be present simultaneously to keep the transverse shortening uniform along the sample's length?

An obvious question concerns a sample that at the outset is homogeneous in chemistry: can such a sample be made inhomogeneous by squeezing it nonuniformly? Yes it can, especially where the two components differ greatly in mobility; an example is provided by the classic experiments on sodium glass by Weber and Goldstein (1964). In the most oversimplified way, we can ignore electrical charge effects and treat the glass as comprising two components, the sodium atoms and the silicate host or matrix. Here the former are vastly more mobile than the latter: in eqn. (14.9b), the joint viscosity N^* is dominated by N^{host}, whereas the joint mobility K^* is dominated by K^{Na}. Because N^* is so large, the sample can be subjected to large stresses and stress gradients while retaining recognizable shape; at the same time, because K^* is large, diffusive movements are rapid enough and extensive enough to be detected experimentally.

In theory, if such an experiment could be sustained, one would observe that, as chemical inhomogeneity builds up, the diffusion rate diminishes; again in theory, one could extrapolate to the state where the diffusion rate becomes zero. At this stage, the concentration profile would be related to the stress profile as in eqn. (14.9b); that is to say, we should have reached the balanced condition that that equation describes (but to actually *reach* the condition would take infinite time).

In more practical terms, eqn. (14.9b) or (14.8) gives an *upper limit* to the composition fluctuation that a stress fluctuation can produce. To go to a more extreme case of contrast in mobility, consider carbon in steel: $1/N^*$ is effectively zero and the limit we seek is given by simply balancing a stress gradient against a composition gradient. Going back to eqn. (14.7a), we need

$$s V^A = \frac{RTc}{C^A} \qquad (14.10)$$

In orders of magnitude, $V^A = 0.012\ \mathrm{m}^3/\mathrm{kg\text{-}mole}$, $R = 8300\ \mathrm{J/kg\text{-}mole}$, $T = 1000\ \mathrm{K}$, $C^A = 0.04$; a stress fluctuation of $100\ \mathrm{MPa}$ gives at most a composition fluctuation of 0.006 or 15% of the amount present. The elastic moduli of steel are of the order of magnitude of $10^5\ \mathrm{MPa}$, so that the stress fluctuation mentioned would give a strain of about 0.001. If the carbon content is homogeneous when the stress fluctuation is first imposed, the composition fluctuation reaches 63% of its ultimate magnitude, or $(1 - 1/e)$, in $10^{13}/4j^2$ seconds. For $j = 1000$ and a fluctuation wavelength of 2π mm, the time is about one month.

The discussion just given has strayed from the central topic of the book; it treats chemical change in *nondeforming* materials, but it serves the purpose of showing that a stress fluctuation takes a long time to produce a fluctuation in composition. If the two components are less extremely contrasted in behavior, the amount of strain that accumulates gets to be large while the composition fluctuation is only beginning to develop. Thus, in practical terms, it is much easier to establish a fluctuation in composition and measure a consequent fluctuation in stress than to work the other way round.

Summary

The material discussed has two components and fluctuates in composition along one direction (taken as x), but is uniform in composition in planes normal to x. The two components interdiffuse, subject to the constraint of no change of dimension in yz planes. If the fluctuation in composition is small and sinusoidal (Figure 14.1), the accompanying fluctuation in transverse compressive stress is also sinusoidal (Figure 14.3). The amplitudes of the two fluctuations are in a ratio fixed by the material properties:

$$ s = \frac{cE^*}{K^* + (1/2j^2N^*)} \qquad \text{as in eqn. (14.9)} $$

E^* is an expression of the *difference* between the two components as regards mobility and molar volume, whereas N^* and K^* express their joint mobility, for change of shape and for diffusion from point to point, respectively. The factor j is the wavenumber of the fluctuation; for sufficiently short waves (large j), the term containing N^* becomes small. This means that the material's ability to change shape at a site becomes of less importance in comparison with the diffusive motions of materials to and fro along the x-direction: when the source regions and sink regions are sufficiently close, diffusion is the dominant material response to a stress fluctuation as well as to fluctuation in composition.

Equation (14.9) describes a series of compatible pairs of values for the stress fluctuation $s \sin jx$ and the composition fluctuation $c \sin jx$. Once the equation is satisfied, the two fluctuations remain in fixed proportion and decay toward zero together.

The condition (14.9) is, in practice, more easily attained by controlling the composition and letting the stress state develop; to establish the stress state and let the composition fluctuation develop is, for most materials, technically more awkward.

If the two components are highly contrasted in both concentration and mobility, as with carbon in steel, N^* is dominated by the less mobile component and K^* is dominated by the more mobile. As in the short-wave extreme case already discussed, the N^* term becomes insignificant in comparison with the diffusive terms.

When the two components are not highly contrasted, it is not obvious how the joint properties N^* and K^* are related to the separate component mobilities. When eqn. (14.9) is written $cE^* = s(K^* + 1/2j^2N^*)$, the left-hand side requires us to imagine separate atoms of the two components, whereas the right-hand side requires us to imagine the assembly of atoms having continuum properties. We wish to assign mobilities whose difference would enter E^* and whose sum or weighted mean value would enter K^* and N^*, but because the underlying concept on the left is different from that on the right, it is not obvious that mobilities can ever be assigned with success. This uncertainty is clarified by turning from binary mixtures to compounds of the type (A, B)X as in Chapter 15.

15

COMPOUNDS OF THE TYPE (A, B)X

Compounds such as gallium aluminum arsenide are of interest for two reasons. First, they have practical use, so that benefit comes from understanding their properties and behavior; but second, they can be regarded as mixtures of just two components, GaAs and AlAs, and so serve as a fresh set of examples of the ideas in Chapter 14. A difficulty in that chapter arises from the fact that at some points we need to imagine a mechanical continuum, while at other points we need to imagine particles traveling independently. In this chapter we need to do the same two things but the conflict in our concepts is not as acute; we can use eqn. (14.9) with more confidence and escape from the sense of an internal contradiction.

The basis for the discussion is the idea of perfect stoichiometry; in a compound of type (A, B)X it is assumed that although the abundance-ratio of A to B is variable, the ratio of (A + B) to X is always exactly 1. Departures from stoichiometry are, of course, of great importance but constitute a later topic.

Theory

We begin in the manner of eqns. (14.7), as if the three species A, B, and X led lives of their own:

$$e^{A} = \frac{s \sin \theta}{6N^{A}} + D^{A}V^{A}j^{2}s \sin \theta - RT\frac{D^{A}}{C^{A}} j^{2}c \sin \theta \qquad (15.1a)$$

$$e^{B} = \frac{s \sin \theta}{6N^{B}} + D^{B}V^{B}j^{2}s \sin \theta + RT\frac{D^{B}}{C^{B}} j^{2}c \sin \theta \qquad (15.1b)$$

$$e^{X} = \frac{s \sin \theta}{6N^{X}} + D^{X}V^{X}j^{2}s \sin \theta \qquad (15.1c)$$

Let the volume-fractions be a, b, and x with $a + b + x = 1$; then in the same tentative manner as before, we propose that $ae^{A} + be^{B} + xe^{X} = 0$ and group the terms from the right-hand sides together:

$$\frac{s \sin \theta}{6}\left(\frac{a}{N^{A}} + \frac{b}{N^{B}} + \frac{x}{N^{X}}\right)$$

$$j^{2}s \sin \theta(aD^{A}V^{A} + bD^{B}V^{B} + xD^{X}V^{X})$$

$$RTj^{2}c \sin \theta\left(b\frac{D^{B}}{C^{B}} - a\frac{D^{A}}{C^{A}}\right)$$

141

As before, the third group shows the tendency of A and B to interdiffuse because of spatial variation in C^A, and the first group shows the material's ability to accommodate a part of the interdiffusion by changing shape; the middle group is more complicated and is discussed last.

As regards change of shape, it is natural to suppose that the three components go along together. They behave as in Figure 14.4c more than as in 14.4b, and if we rewrite the group as $s \sin \theta / 6N^*$, then, as in Chapter 14, we lean more toward the form $N^* = (aN^A + bN^B + xN^X)$, where all three components have similar velocity fields.

As regards change of composition, the situation is even more constrained and symmetrical than in Chapter 14. Not only is dC^B/dx the exact opposite of dC^A/dx, but also, because of the stoichiometry, the number of atoms of A that cross any plane of X atoms must be balanced by an equal number of atoms of B crossing the other way; the *number-fluxes* must be equal and opposite, when measured in a frame attached to the substrate of X. Even if B is intrinsically more mobile than A, the stoichiometry hinders B and stimulates A until a single value for the number-flux describes both motions.

This fact permits an act of the imagination that is a help when we turn to the middle group; we imagine that an atom of B can be turned into an atom of A by adding a modifier a. In place of, say, 100 atoms comprising 63 of B and 37 of A, we imagine 100 atoms of B, of which 37 have the modifier attached to them, as in Figure 15.1b. The main properties of a are its volume—we make $V^a = V^{AX} - V^{BX}$, and its mobility—we assume a to be driven by dC^A/dx with the same number-flux as affects A and B in the earlier description. But, with a, there is only one flux; the entire substrate consists of BX everywhere and cannot change its own composition; it can only act as host to a variable concentration of additive a. Within the substrate, there is a population of a with nonuniform concentration C^A, that tends to smooth itself out; and because a has volume V^a, the motion of a requires compatible motions of the substrate to satisfy $e_{zz}^{\text{total}} = 0$.

The description now consists of just two components, the substrate BX and the free-floating additives a, as in Figure 15.1c. It is now apparent that the middle group of terms must separate into two parts. The term as a whole represents the fact that, if at some site $\partial^2 \sigma_{zz}/\partial x^2$ is not zero, material will be flowing to the site or away from it by diffusive motions. If the material consists of a substrate BX and an additive a, the group must reduce to

$$j^2 s \sin \theta (a^a D^a V^a + b D^{BX} V^{BX})$$

Here a^a is the volume fraction of the additive; a^a is much less than b and b is close to 1. The act of imagining the additive a enables us to satisfy the stoichiometry constraint automatically and at the same time reduce from three terms to two. (As an example, consider (Fe, Zn)S. Molar volumes are, for ZnS, 0.0238 m^3/kg-mol and for FeS, 0.0246 m^3/kg-mol, with difference 0.0008 m^3/kg-mol. Assuming linearity, 63 mol of ZnS and 37 mol of FeS would occupy 2.41 m^3. In the alternative form of description, this would be

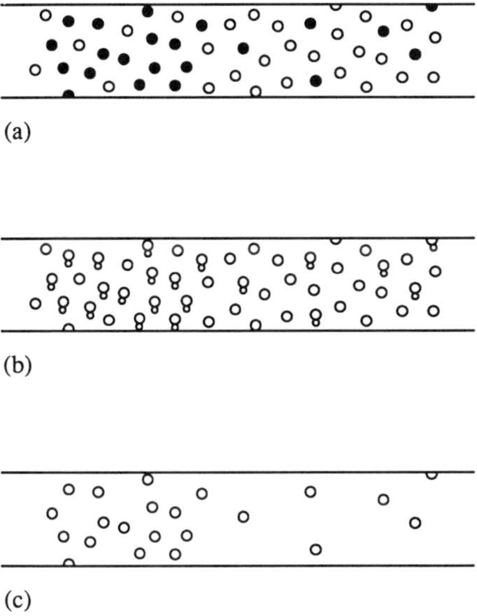

(a)

(b)

(c)

Figure 15.1 Three representations of a nonuniform binary material of formula (A, B)X. (a) Filled circles show atoms of component A and open circles show component B; component X is not shown. (b) Atoms of A are conceived as atoms of B to which an additive has been attached. (c) The material is conceived as a uniform substrate of BX, on which a non-uniform distribution of the additive is imposed.

viewed as 100 mol of ZnS, or 2.38 m^3, plus 37 mol of additive with volume 37×0.0008 or 0.03 m^3. Thus volume fractions are $b = 0.99$ and $a^a = 0.012$.)

The relation of the a-term and the BX term to each other, and to the first and third groups already discussed, is more clear-cut than in Chapter 14. The term $D^{BX}V^{BX}$ shows the self-diffusion of the whole assembly away from high-compression sites; if, as elsewhere, we use $K^{BX} = 3D^{BX}V^{BX}$, we envisage a characteristic length for the total material given by $2(N^*K^{BX})^{1/2}$, and this length is related to the mechanism, whatever it may be, by which the mass of X atoms changes dimension. By contrast, the term D^aV^a does not involve X at all; it is related to the quite different mechanism by which an atom of A and an atom of B exchange places; this part of the middle term is more closely linked to the third group, the composition-driven interdiffusion.

To pursue the resemblance just mentioned, we recall the stoichiometry constraint. Because of this constraint, the third group is better written as $a^aRTj^2c \sin \theta(D^a/C^a)$; as already discussed, the separate mobilities D^A and D^B are forced to amalgamate to the single mobility D^a. Then this group and part of the middle group can be written as $j^2(\sin \theta)a^aD^a[sV^a + cRT/C^a]$. This combination describes the interdiffusion of A and B (using the fictitious additive a). The single mobility D^a controls the flux, but the driving force for interdiffusion contains two parts, the stress fluctuation and the composition fluctuation. The second is invariable, but the first acts only because of the volume-difference V^a; it is proportional to this factor, and would be zero if two atoms of the same size substituted for each other.

The description in terms of a *substrate that self-diffuses* plus *components that interdiffuse* permits a further distinction: it is the substrate that has continuum properties, to which the reasoning in Chapter 11 applies, and for which we use eqn. (12.7) specifically along one direction or another. The interdiffusive effects, here mimicked by the motion of the additive a, do not resemble continuum behavior; in isotropic materials, the additive a affects only the volume of a sample-element and cannot affect its shape; the additive responds directly to the mean stress and produces only an isotropic change in mean strain. In the cylinder problem treated above, the symmetry and uniformity assumed are such that this distinction leaves the mathematical solution unchanged in form. But if a less regular physical situation were to be treated, the distinction between the behavior of BX and the behavior of the additive would have more noticeable consequences.

Review. Equation (14.9) contains a dilemma: when written in the form $cE^* = s(K^* + 1/2j^2N^*)$, it seems to require that we consider the same material (a binary mixture of A and B) as a mass of separate particles on the left-hand side and as a mechanical continuum on the right-hand side. One can imagine the two behaviors in a single material easily enough, as in Chapter 13; the difficulty arises when we try to assign the components A and B individual mobilities and then form joint properties. If we form E^* using the separate-particles picture and K^* using the continuum picture, and then put the E^* and K^* into one equation, it is hard to see how the equation can be a good description of any real behavior.

With the preceding ideas in mind, the amalgamation of the three equations, set (15.1), can be tried again. On the left-hand side, we seek a total strain of zero; on the right, we form three terms as follows (dropping $\sin\theta$):

$$0 = \frac{s}{6N^*} + j^2s(bD^{BX}V^{BX} + a^aD^aV^a) - RTj^2ca^a\frac{D^a}{C^a} \tag{15.2a}$$

or

$$s = \left[\frac{2j^2E^a}{(1/N^*) + 2j^2(bK^{BX} + a^aK^a)}\right]c \tag{15.2b}$$

Here $E^a = RTK^a/V^{BX}$. The procedure is similar to that used for eqn. (14.9), but in that discussion the two diffusivities K^A and K^B were of the same type, whereas in eqn. (15.2b) they are definitely different: K^{BX} is a joint diffusivity, a self-diffusivity for the ensemble (A, B)X, which is presumably dominated by the self-diffusivity of the substrate species X and can lead to local change of shape; but K^a is wholly independent of species X and describes only the ease with which A and B are caused to interdiffuse by nonuniform pressure, leading only to local change of volume. Because V^{BX} and K^{BX} are properties of the entire substrate moving en masse, they parallel N^* and are hereafter designated V^* and K^*.

In the same way that (14.9) might be called the key equation of Chapter 14, so (15.2) is the key equation of this chapter. In the remaining pages, discussion revolves around this equation. A small extension leads to one additional term and to eqn. (15.6), but the main *idea* of the chapter has just been stated—that to treat the material as substrate plus additive adds definiteness. The function of the remaining pages is to enlarge upon the previous paragraph.

Orders of magnitude. Compounds of the type (A, B)X have the advantage, as just discussed, that one can distinguish readily between diffusion of the ensemble (all components traveling in the same direction) and interdiffusion (A traveling one way and B the other.) But they have the disadvantage that their creep behavior is, for the most part, non-linear with the driving stress magnitude. The theory assumes linear behavior, and stress/creep-rate measurements that are even approximately compatible with that assumption are scarce.

For the sake of orders of magnitude, estimates for the mixed orthosilicate *olivine* can be used. The formula is $(Mg, Fe)_2SiO_4$. The oxygen atoms are the main space-filling component and the creep viscosity, roughly 10^{18} Pa-sec at 1700 K, is attributed to movement of these as single atoms. Hence, for present purposes, X is a single oxygen atom, and A and B have to be given volumes appropriate to half a magnesium atom or half an iron atom; the silicon atoms, of course, play a large role in fixing the oxygen mobility but do not contribute directly to any arithmetic term in eqn. (15.2).

On this basis, needed quantities are as follows:

N^*	viscosity of the $(Mg, Fe)_2SiO_4$ ensemble, 10^{18} Pa-sec
	volume of 1 kg-mole of Fe_2SiO_4, 52×10^{-3} m^3/kg-mol
	volume of 1 kg-mole of Mg_2SiO_4, 50×10^{-3} m^3/kg-mol
	volume-difference Fe − Mg, 2×10^{-3} m^3/kg-mol
V^*	volume of 1 kg-atom of oxygen, 13×10^{-3}/kg-atom
V^a	volume-difference Fe − Mg per kg-atom of oxygen, 5×10^{-4} m^3/kg-atom
D^i_{oxy}	isotope self-diffusion coefficient for oxygen atoms, 10^{-18} m^2/sec
D^i_{FM}	isotope self-diffusion coefficient for iron or magnesium atoms, 3×10^{-14} m^2/sec
R	gas constant, 8×10^3 J/kg-atom-K
C^a	atom abundance Fe/(Fe + Mg) (an arbitrary typical value), 0.25
K^*	$D^i V^*/RT$ for oxygen atoms, 10^{-27} m^2/Pa-sec
K^a	$D^i V^a/RT$ for motion of "component a," i.e., for exchange of Fe for Mg, 10^{-24} m^2/Pa-sec
E^a	$D^i_{FM} V^a/V^*$, 10^{-15} m^2/Pa-sec
a^a	volume fraction of additive, $C^a V^a/V^*$, 10^{-2}
b	volume fraction of substrate, 1

(Diffusivities are from Freer (1981).)

In this example, the sum $bK^* + a^aK^a$ is dominated by a^aK^a, $= 10^{-26}$ m^2/Pa-sec, so that

$$s = \left[\frac{10^{-15}}{(10^{-18}/2j^2) + 10^{-26}}\right]c$$

The terms in the denominator are of the same order of magnitude when $j = 10^4$ and the wavelength λ is about 0.6 mm. At this scale, if c is 1% or 0.01, s is 10^9 Pa or 10 kilobars: at such a high temperature, for components of comparatively high mobility that are only slightly different in effective volume, very large fluctuations in stress are needed to stabilize small fluctuations in composition. If the material were more viscous, or j were larger, the N^* term would become insignificant and the relation of s to c would be hardly affected; but if j is made smaller, e.g., $j = 100$ and $\lambda = 6$ cm, the interdiffusive tendency is weakened, it is accommodated mostly by change of shape of the oxygen substrate, and s drops to about 1 bar (10^5 Pa).

Further details brought to light by this numerical example are:

1. The sharp transition at around $j = 10^4$; for smaller j-values, s is proportional to j^2, whereas for larger j-values, s reaches a plateau. See Figure 15.2; on the plateau, relations are as they were for carbon and steel, already discussed.
2. The small part played by the pressure-driven self-diffusion of the oxygen atoms (the K^* term); where K^* and K^a are so different, the K^*-contribution is negligible. But in a true binary mixture, the disparity would be smaller and the K^* term would be of more consequence.
3. The limited scope for lowering the stress at which the plateau appears. To work at lower temperature gives a lower plateau, as in the carbon–steel example; s_{plateau} would be reduced by a factor of 1/6 at room temperature. Or one might find a system where one or other of the relevant volumes was larger—either the unit volume of the host X or the difference in

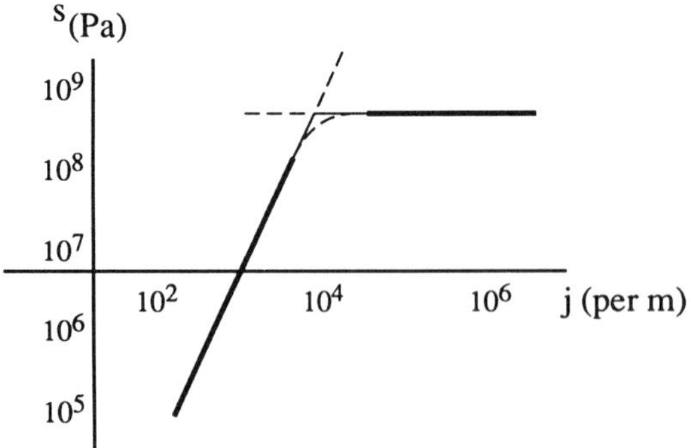

Figure 15.2 Variation of the stress increment s with wavenumber j.

volume V^a between the unit volumes V^{AX} and V^{BX}. But, short of turning attention to molecular mixtures, one cannot diminish s_{plateau} very much, hardly below 10^8 Pa.

4. The limited scope for raising the value of j at the change of slope. The critical value of $2j^2$ is $1/N^*(bK^* + a^aK^a)$. Reducing K^a becomes ineffective as soon as a^aK^a is as small as bK^*, and reducing K^* is ineffective because any change that reduces K^* tends to increase N^*. In fact, $N^*K^* \sim L^2$, the characteristic length for the host or substrate species; the only way to change to a larger j-value is to change to a system where the mechanism for change of shape involves smaller displacements of atoms in space. In the olivine system discussed, the shape changes by dislocation-climb, with oxygen atoms migrating along dislocation loops for distances of the order of a micrometer or more; by contrast, in a deforming hot metal, at least part of the deformation is accomplished by movements on nanometer scale, with j correspondingly one thousand times larger. The field where plateau conditions would be encountered in real life becomes very narrow and highly specialized.

A nonzero constriction rate

Up to this point, the interaction of chemistry and deformation—as embodied in eqn. (15.2)—has been discussed wholly for the condition of no change in material dimensions, $e_{zz}(\text{total}) = 0$. To make the discussion more widely applicable, and as a step toward considering interfaces, we now consider the possibility that e_{zz} has a nonzero value, while remaining uniform at all points in the sample. Groundwork was laid in Appendix 12B, where the idea was introduced of a material whose viscosity varies along the x-direction as $N_0 + n \sin jx$.

In place of eqn. (14.2) where the mean value of σ_{zz} is equal to σ_{xx}, we propose

$$\sigma_{zz} = \sigma_{xx} + S_0 + s \sin jx \tag{15.3}$$

Equation (15.1a) becomes

$$e^A = \frac{S_0}{6N^A} + \frac{s \sin jx}{6N^A} + \text{remaining terms as before} \tag{15.4}$$

Equations (15.1b) and (15.1c) are modified in a similar way; then if the prescribed uniform value of e_{zz} is e_0, and we seek the condition $e_0 = ae^A + be^B + xe^X$, the total left-hand side becomes e_0 and the total right-hand side is as before except for the extra terms with S_0, which total $S_0/6N^*$. If $N^* = N_0 + n \sin jx$ and n is small with respect to N_0,

$$\frac{S_0}{6N^*} = \frac{S_0}{6N_0} - \frac{S_0 n}{6N_0^2} \sin jx \tag{15.5}$$

It follows from eqn. (15.5) that for the strain rate to be uniform with respect to x, with magnitude e_0, we need first $e_0 = S_0/6N_0$ and, second, the sum of all the terms containing the factor $\sin jx$ must be zero. They are as already shown in eqn. (15.2a) with the addition of the extra term

$$-\frac{S_0 n}{6N_0^2} \sin jx \qquad \text{or} \qquad -e_0 \frac{n}{N_0} \sin jx$$

Equation (15.2a) becomes

$$\frac{s}{6N^*} + j^2 s(bD^{BX}V^{BX} + a^a D^a V^a) - RTj^2 cD^a - e_0 \frac{n}{N_0} = 0 \qquad (15.6a)$$

A situation of particular interest is one where the supposed fluctuation n in viscosity is linked to the fluctuation c in composition; from the nature of N^*, such a link seems unavoidable. If we express the link by writing $n = kcN_0$, eqn. (15.2b) becomes

$$s = \left[\frac{2j^2 E^a + 6e_0 k}{(1/N^*) + 2j^2(bK^* + a^a K^a)} \right] c \qquad (15.6b)$$

The factor k can be positive or negative and shows how much variation in the material's viscosity is produced by a given small variation in its composition.

The equation shows more fully than (15.2) how nonuniform composition and nonuniform stress are interlinked. The three terms in the denominator show, as already described, the three means by which stress affects a sample's dimensions: by changing its shape, by causing self-diffusion of the ensemble away from highly compressed sites, and by driving interdiffusion of the exchangeable components A and B. The two terms in the numerator show the two ways in which nonuniform composition produces nonuniform stress—(i) directly, by causing interdiffusion that causes changes of volume, and (ii) indirectly, by modifying the material's viscosity, so that even a uniform imposed strain rate engenders a stress variation.

Binary materials reconsidered

Equations (15.2b) and (15.6b) show a denominator containing the term $2j^2(bK^* + a^a K^a)$, where eqn. (14.9b) has only $2j^2 K^*$. The separateness of K^* and K^a in this chapter sheds light on the question of what K^* actually represents in Chapter 14—a question that was left in an unsatisfactory state in that chapter. The purpose of this section is to review ideas about forming K^* from K^A and K^B in a simple binary system.

The picture from which K^* and K^a emerge is the mixed compound (A, B)X, on which a stress fluctuation has two effects. Consider a site where the compressive stress is at a local maximum, with a local minimum or sink-region somewhere nearby. The two stress-driven diffusion effects are

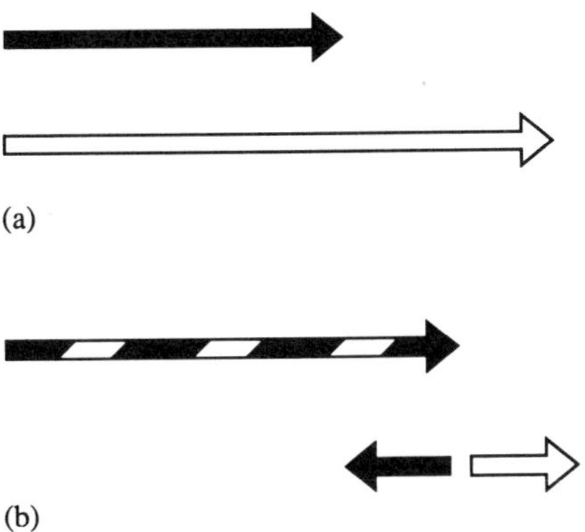

(a)

(b)

Figure 15.3 Two separate velocity fields as in (a) can be treated as the superposition of a joint velocity field and a velocity field for interpenetration, as in (b).

(i) self-diffusion of the whole ensemble from source to sink and (ii) inter-diffusion of components A and B, with whichever occupies less space migrating *toward* the high-stress site because of the stoichiometry constraint. The first effect, where the difference between components A and B is immaterial, is as discussed in Appendix 12B and Chapter 13. The second is linked to the stoichiometry constraint, which permits the treatment in terms of the single imagined additive constituent a, as in Figure 15.1c. We now ask whether two effects as distinct as these two might be present in a simple binary material.

The answer is that two comparable effects can certainly be separated by an algebraic step or trick; but the step is more than *merely* a trick, and does indeed resolve some of the uncertainty as to whether a material can be represented both as a continuum and as particles (see Chapter 14, last paragraph of the Summary).

The algebraic step is simple: rather than giving attention to A and B separately and imagining a velocity-field for A and a separate velocity-field for B as in Figures 14.2 and 15.3a, we conceive a joint or average velocity-field for A and B together, plus a second velocity field that shows their differences, as in Figure 15.3b. The joint field is selected so that what remains to be superimposed on it to produce the right total displacements or flows is a small adjustment for component A and an *exactly equal and opposite* small adjustment for component B. The value of this format is that the needed material-mobility factors are more easily identified: the joint velocity field will be governed by a K-property like K^* discussed above, that is closely related to N^*; for all of these factors, the material can quite reasonably be conceived as a continuum. By contrast, for the opposing flows, counterflows or interpenetrating parts, the exchange of A and B, there must be a single K-property like K^a discussed above that is closely related to E^a or E^*; for all these factors, the material can reasonably be conceived as particles. No stoichiometry constraint actually operates, but by describing a binary material in the manner of Figure 15.3b, one sees that it is likely to behave *as if* it were an (A, B)X compound.

The conclusion is that even for a simple binary material, we should not look for a *single* factor K^* constructed from K^A and K^B as in eqn. (14.9). Instead, we should construct two factors, K^{joint} and K^{inter}, and write the denominator in eqn. (14.9b) as $1/N^* + 2j^2(pK^{joint} + qK^{inter})$, where p and q are suitable weighting factors based on the components' unit volumes, as in the example on page 145. Here the numbers are set aside to avoid obscuring the following suggestion: the use, in one equation, of two groups of terms—one group derived by treating a material as atoms and the other group derived by treating the material as a continuum—does not invalidate the equation. But each group should be comprehensive and no term should appear that does not clearly belong in one group or the other. The replacing of K^* from Chapter 14 by $pK^{joint} + qK^{inter}$ is an attempt to follow this precept.

Summary

The behavior of a material of composition (A, B)X contains two parts (i) behavior shown by the ensemble and (ii) behavior that involves atoms of A exchanging position with atoms of B.

The ensemble can be treated as a material of fixed composition. It shows the two behaviors described in Chapter 13, namely, change of shape at constant volume and change of volume at constant density by diffusive movement of material from high-compression sites to sites of lower compression.

The interdiffusive exchange also has two parts, one due to nonuniformity of stress and the other due to nonuniformity of composition.

The interdiffusive effects can be described by means of an imagined additive a. Regarding volume, $V^a = V^A - V^B$ and similarly for other properties. Then an interchange of an atom of A and an atom of B can be described as a single motion of a particle of additive from one site to another.

The four right-hand-side terms in eqn. (15.2a) correspond to the four behaviors distinguished:

$s/6N^*$	change of shape of the ensemble, components A and B deforming together
$j^2sbD^{BX}V^{BX}$	stress-driven self-diffusion of the ensemble, components A and B traveling together
$j^2sa^aD^aV^a$	stress-driven motion of the imaginary additive ($=$stress-driven interchange of A and B)
$RTj^2ca^aD^a/C^a$	composition-driven motion of the additive ($=$composition-driven interchange of A and B)

The first two behaviors affect different directions in the material to different extents; they are driven by separate components of the stress tensor according to eqn. (12.7). By contrast, the last two behaviors lead to isotropic change of volume, and it is variation in the mean stress that enters the third term.

The preceding statements describe behavior in a condition of cylindrical symmetry about one direction (x), uniformity in all planes normal to x, and no strain along all lines normal to x. In a condition of uniform constriction (strain rate = an arbitrary rate e_0 along all lines normal to x), any variation in viscosity due to variation in composition will give rise to a fifth effect as in eqn. (15.6).

16

TWO PHASES AND TWO COMPONENTS

The overall plan of Chapters 13 through 16 is shown in the introductory figure of Part III. As the figure suggests, the objective of this chapter can be stated in three ways:

  (i) to extend ideas from Chapter 15 to a situation where an interface is present, or
 (ii) to extend ideas from Chapter 13 to a situation where the material envisaged is of variable composition, or
(iii) to combine the main ideas from Chapter 14 and 15 with the main ideas from Chapter 13.

The program of the present chapter is to do a little of (i) and then a little of (ii), before attempting a total union.

By way of preview, it is to be noted that an interface may move with respect to the material on either side or, in other words, material may change phase, crossing the interface as it does so. Situations with a moving interface are more complicated than situations where the interface is stationary. In both part (i) and part (ii) of this chapter, the fact that the interface moves is recognized but disregarded. The reason is that in part (iii), a particular combination of circumstances is examined that leads to no movement of the interface. Such a combination is highly special and not likely to be met in the real world, but it shows particularly clearly how chemical change and a material's deformation interact. The overall program then is to lead up to discussion of this stationary-interface state.

Exponential change of composition and stress

Chapters 14 and 15 emphasize spatial variations of a harmonic kind, continuing through space at constant amplitude to an unlimited extent. The essential difference in this chapter is that here we treat effects that have maximum intensity at or close to an interface, and die away to insignificance in the remote interiors of the two touching phases. Exponential functions were found useful to describe an effect of this type in Chapter 13, and it is natural to seek a physical situation where they can be used again; the first step is to distinguish the two types of behavior in Figure 16.1.

The purpose of the figure is to emphasize behavior as in diagram (a). In

152

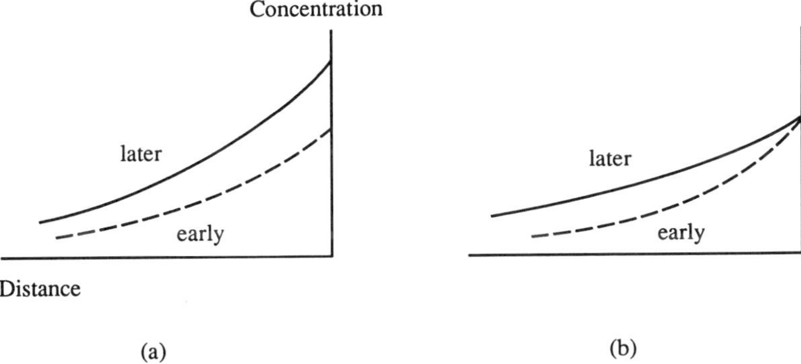

Figure 16.1 (a) Change of concentration in the interior of a sample where the concentration at the boundary increases with time; if the increase is exponential with time, the profiles can be exponential through space. (b) Change of concentration where the concentration at the boundary is jumped to a new value and then kept constant; the profiles are of error-function type.

experimental work, behavior as in diagram (b) is commonly explored: conditions are established at the interface that are different from conditions in the sample's interior (in respect of composition or temperature or other variable); then, as time passes, the interface state is kept steady while the gradient into the interior gradually flattens. But in this situation the profiles are not exponential curves, and in other ways as well the behavior in diagram (a) is a closer parallel with behavior in Chapter 15.

The essential feature of diagram (a) is that the full line is an exponential curve, $y = Ae^{kx}$ or $y = Ae^{x/B}$, where A is its height at $x = 0$ and B or $1/k$ is the distance along x over which the height diminishes to A/e. For a moment, we leave stresses out of consideration and imagine that some diffusing component has a concentration profile as shown by the full curve; then the concentration at any point must currently be increasing at a rate given by

$$\frac{\partial c}{\partial t} = (\text{mobility coefficient}) \cdot \frac{\partial^2 c}{\partial x^2} \qquad (16.1)$$

[cf. eqns. (3.4) and (8.6)], and this must be true at $x = 0$ just as much as at any other point: the profile extending into the interior is a simple exponential curve only if the boundary value is continually increasing, and indeed increasing at an exponential rate. Clearly this is an evolution that we want to consider only in its early stages, when change at the interface is occurring at some believable rate.

Turning to stress state, it is recalled that the main ideas of Chapters 14 and 15 are embodied in the four terms in eqn. (15.2); these four terms show how the stress fluctuation s and the composition fluctuation c interact. In fact the equation identifies a steady state—a relation between s and c that is able to persist through time while the magnitudes of s and c both change. If a parallel situation can exist at an interface, and there the concentration is slowly increasing with time, the stress must necessarily increase with time as well. For a physical situation that could show such behavior, see Figure 16.2.

Figure 16.2a shows in diagram form two phases both having composition (A, B)X. When they coexist in equilibrium, phase g has a higher ratio of A to B, as in diagrams (a) and (b). In a particular sample of bulk composition Z, the two phases have compositions g_1 and h_1 at temperature T_1; then if the temperature is moved suddenly to T_2, several changes occur: right at the interface, a small quantity of material changes from phase g to phase h; in this conversion, atoms of A are liberated and atoms of B are absorbed; the result is a thin layer of phase g with composition g_2 touching a thin layer of phase h with composition h_2, as in Figure 16.2c. These layers are open to diffusive exchange with material in the phase-interiors; atoms of A diffuse away from the interface in both directions; and if temperature T_2 were maintained, composition profiles as in Figure 16.2d would develop; that is, the evolution would resemble that shown in Figure 16.1b. Compositions at the interface would remain at g_2 and h_2, with material changing phase at just a sufficient rate to match the diffusive processes.

For the purposes of this chapter, however, we do not suppose that the temperature is jumped to T_2 and then kept constant; instead we suppose that the temperature is raised in a slow, continuous way. The composition at the interface follows the temperature, always failing quite to attain equilibrium but departing from equilibrium only by a small amount, which we shall for the most part ignore. The evolution then resembles more the one illustrated in Figure 16.1a and, assuming as in Chapter 15 that there is a small volume-difference $V^a = V^A - V^B$, the diffusion of A from the interface into the two interiors will be accompanied by a nonuniform stress effect σ_{zz}. (Axes are used as in Chapter 13 with x normal to the interface, cylindrical symmetry about x, all quantities uniform along y and z, and the strain rate e_{zz} equal to zero everywhere.)

The situation described is intentionally very similar to that in Chapter 15. The length scale B or $1/k$ of the profiles is controlled by the rate of increase of temperature; the slower the rate of increase, the longer the length scale B. Then the four effects that combine to satisfy the overall constraint $e_{zz} = 0$ are exactly the same as in eqn. (15.2). The only differences are that now $s \sin jx$ is replaced by se^{kx}, $c \sin jx$ is replaced by ce^{kx}, the common factor that disappears in going from equations like (15.1) to equations like (15.2) is now e^{kx}, and all terms except the term with N^* change sign. The last difference arises from the fact that the second derivative of e^x equals the original function, whereas the second derivative of $\sin x$ equals the negative of the original function.

The result, corresponding to eqn. (15.2b), is that when $e_{zz} = 0$, for a composition profile and a stress profile to steepen together while remaining

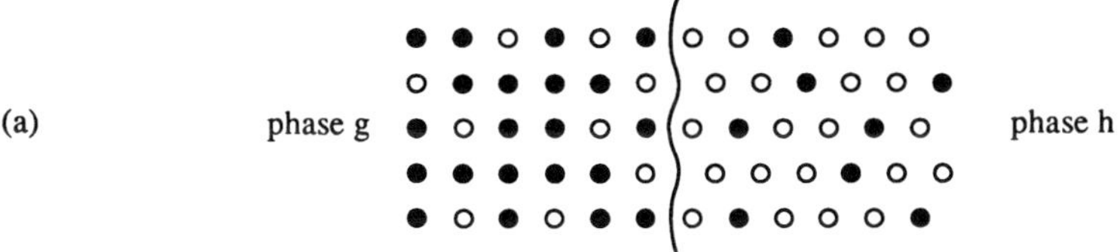

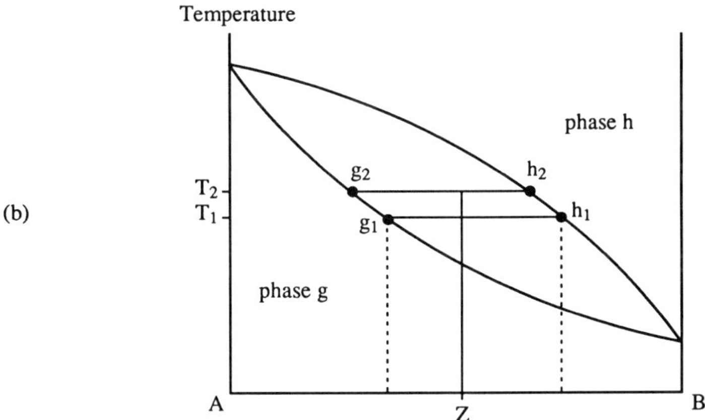

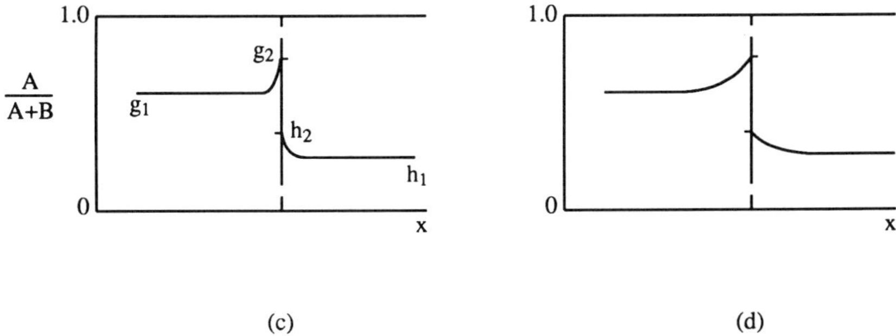

Figure 16.2 A boundary where compositions are disturbed by jumping the temperature. (a) The initial state, a perfectly coherent boundary. (b) The phase relations. (c) Profiles of composition very soon after a jump in temperature. (d) Profiles of composition at a longer time after the jump in temperature.

in a fixed ratio,

$$s = \left[\frac{2k^2 E^{\mathrm{a}}}{(1/N^*) - 2k^2(bK^* + a^{\mathrm{a}}K^{\mathrm{a}})} \right] c \qquad (16.2a)$$

The physical interpretation of the separate terms is as before (see, for example, Chapter 15 Summary) but the negative sign in the denominator requires immediate comment. The equation suggests that for certain magnitudes of the parameters the denominator will be zero, and that a small composition variation ce^{kx} will be accompanied by an infinitely large stress variation. What physical processes lie behind this improbable suggestion?

To clarify the situation, we write eqn. (16.2a) as

$$\frac{s}{N^*} - 2k^2(bK^* + a^{\mathrm{a}}K^{\mathrm{a}})s = 2k^2 E^{\mathrm{a}}c \qquad (16.2b)$$

The right-hand-side shows the rate at which the z-dimension of a material element tends to increase because of composition-driven inflow of the additive a; the left-hand-side first term shows the rate at which the z-dimension tends to decrease because of σ_{zz} exceeding σ_{xx}—the constriction effect, change of shape at constant volume; and the left-hand-side second term shows the rate at which the z-dimension tends to change because of the two stress-driven diffusion effects, first for self-diffusion of the (A, B)X ensemble and second for interdiffusion of A and B, as represented by a one-way flux of the additive a. It is a characteristic of harmonic stress profiles that *change of shape* and *change of volume by diffusion* have similar effects: at a site where σ_{zz} is locally high, both behaviors cooperate to diminish the z-dimension at that site. But in an exponential profile the two behaviors compete: as already noted, the second derivative now has the other sign, and stress-driven diffusion now tends to *increase* the z-dimension of an element at all sites where s is positive.

For sufficiently large values of the length B or small values of k, change of shape prevails. But if the whole temperature-history is programmed to move on a faster time scale, B takes a smaller value, k takes a larger value and a given variation c in composition requires a larger variation in the stress σ_{zz} "to make room for influx of a." In physical terms, for sufficiently rapid change of temperature, the steady state described by eqn. (16.2) cannot be attained. Obviously, stress magnitudes do not become infinite—instead, change of composition lags behind; c does not become as large, in proportion to s, as eqn. (16.2) suggests, and compositions even at the interface itself fail to change as fast as the temperature. In other words, the excess temperature required to drive the change-of-phase process is no longer a negligibly small quantity. However rapidly one changes the interface temperature, one cannot in fact produce an exponential profile of concentration with a parameter k as large as the critical value. It is only smaller values of k, i.e., larger values of the length scale B, that can be produced by the procedure described.

The particular complication just discussed should not obscure the main thrust of the chapter so far, which is that, to a large extent, ideas from Chapters 14 and 15 carry over to situations where an interface is present with little change. But ideas related to a nonzero constriction rate e_{zz} have yet to be considered: to open this topic we go to part (ii) of the program on page 152 using Chapter 13 as basis or starting point.

(A second complication is as follows: the excess of component A that diffuses away from the interface in both directions and in some sense creates the extra stresses is produced by change of phase: the interface must move with respect to the main bulk of the two phases in order for any of the processes discussed above to run. This movement of the interface affects the shape of the profiles on either side. But this complication also can be set aside for the present; the main line of development of the chapter can be followed without pursuing that aspect.)

A nonzero constriction rate

The situation imagined is as in Chapter 13: a planar interface that is taken as the plane $x = 0$, all quantities uniform along directions y and z, and cylindrical symmetry about any line parallel to x. In Chapter 13 the materials were chemically invariant; here we switch to supposing that the material in the space $x < 0$ has composition (A, B)X. A uniform constriction strain rate e_{zz} is imposed, with magnitude e_0, and we enquire whether any composition profile of A along x will be steady as deformation proceeds. As in Chapter 13, one may assume that the interface is ideally coherent, with a second polymorph also (A, B)X occupying the space $x > 0$ (see Figure 16.2a); or, to represent an interface that is *not* coherent, one may assume an ensemble of three materials, as in Figure 13.5, where the third material is of such low viscosity that, within it, the stress difference $\sigma_{zz} - \sigma_{xx}$ is negligibly small.

To form expectations about the material composition, we have to consider its stress state. In Chapter 13, it is established that if the initial state is hydrostatic, with compressive stress equal to σ_{xx} everywhere, then introducing the strain rate $e_{zz} = e_0$ has two effects: first, the stress magnitude σ_{zz} takes on an exponential form, and approaches $\sigma_{xx} + 6Ne_0$ at points remote from the interface; second, material begins to diffuse down the stress gradient thus created, and begins to cross the interface—or, in other words, the interface begins to move through the material (Figures 13.2d and e). Both effects bear on the material's composition, but for a start we ignore the second so as to focus on the first.

A situation particularly easy to imagine is as follows: let the space $x < 0$ be occupied by a stiff material (A, B)X; let the plane $x = 0$ be an interface against a material also of formula (A, B)X but of much lower viscosity; but let this material be just a thin layer, bounded at say $x = \Delta$ by another interface, and let the material in the space $x > \Delta$ match the material in the space $x < 0$. Then if atoms of A are larger than atoms of B or the combined

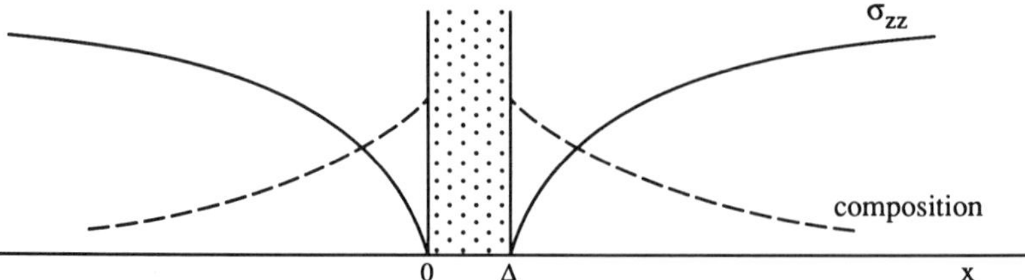

Figure 16.3 Profiles of composition (broken line) and compressive stress σ_{zz} (full line) as they might be on either side of a slab of weak material of thickness Δ.

form AX occupies more space than BX, when the strain rate e_0 is imposed, atoms of A will migrate to the space between zero and Δ, the concentration of A in that space will rise, and component A will stop migrating only when a sufficient concentration gradient has built up to oppose the stress gradient; see Figure 16.3.

As discussed in Chapter 15, the interdiffusion or exchange of atoms of A and B is not a continuum behavior; it is driven strictly by variation in mean stress and results only in isotropic change of volume. From eqn. (13.5a) we have:

$$\sigma_{zz} = \sigma_{xx} + 6N^*e_0 - He^{x/B} \tag{16.3}$$

(The preexponential stress magnitude formerly designated A is here designated H to avoid confusion with component A.) Then the mean stress σ_m is given by

$$\sigma_m = \sigma_{xx} + 2(6N^*e_0 - He^{x/B})/3 \tag{16.4}$$

and the effect on the chemical potential of A is given by

$$\mu^A = \mu_0^A + 2V^A(6N^*e_0 - He^{x/B})/3 \tag{16.5}$$

where μ_0^A is the potential under an isotropic compression p of magnitude equal to σ_{xx}. Then if the composition varies in such a way that

$$RT \ln C^A = RT \ln C_0^A + 2V^A(He^{x/B})/3 \tag{16.6}$$

where C_0^A does not vary with x, there is no net variation of μ^A along x.

We should notice the physical picture behind this equation. A stiff material with a planar interface is suffering constriction; away from the interface in the material's interior, the radial compressive stress that drives constriction is comparatively high, but close to the interface it is lower, the stress profile being exponential (Figure 16.4a). Any component will have a tendency to migrate from high stress to lower stress but it is possible for this tendency to be exactly balanced by a suitable gradient or profile of composition (Figure 16.4b). In eqn. (16.6) the composition is related to H, the stress deficiency at the interface, but H in turn is related to the constriction rate.

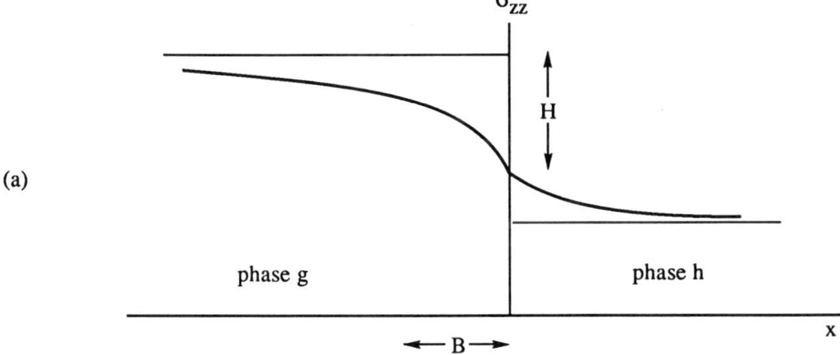

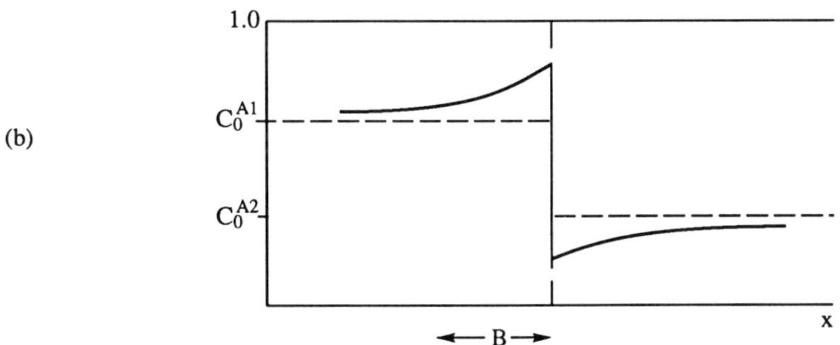

Figure 16.4 (a) Profiles of compressive stress as they might be on either side of an ideally coherent interface. (b) Profiles of composition compatible with the stress profiles in (a).

The equation thus shows that if a planar-bounded sample is subjected to constriction, there is a particular concentration profile that can remain steady as constriction proceeds. The height of this profile is fixed by the constriction rate and its length scale is B, the same as for the stress profile, B being fixed by the behavior of the substrate: $B = (2N^*K^*)^{1/2}$.

We contrast this situation with the one discussed in the previous section. There the constriction rate is zero, an exponential profile of concentration is maintained by *continually changing* the concentration at the interface, and the length scale is fixed by the joint mobility of the interdiffusing species A and B.

Details. There are three points to be noted in connection with eqn. (16.6). First, it is subject to the same limitation as earlier parts of this chapter: it is not exact because it contains no allowance for the fact that the interface moves. As before, this limitation is ignored because the intention is to lead up to a situation where the interface does not move.

Second, the equation can be simplified if the variation in composition is small. The equation envisages an actual composition C^A at any point along x, and a reference composition C_0^A. Let the difference be designated c_x. Then if c_x/C_0 is small enough to permit the approximation $\ln(1 + c_x/C_0) = c_x/C_0$, eqn. (16.6) becomes

$$c_x = \frac{2V^A C_0^A}{3RT} H e^{x/B} \tag{16.7a}$$

Further, let c designate the maximum value of the variation in composition, at the point $x = 0$. Then

$$c = \frac{2V^A C_0^A}{3RT} H \tag{16.7b}$$

where H is the maximum value of the variation in stress σ_{zz}.

The corresponding result for exchange of two components is

$$c = \frac{2V^a C_0^a}{3RT} H \tag{16.7c}$$

For example, if $V^a = 0.001$ m^3/kg-mol, $C_0^a = 0.25$, $H = 80$ MPa, $R = 8300$ J/kg-mol-K, and $T = 1000$ K, then $c = 0.002$ or 0.2%.

The third point is that, for purposes of introducing eqn. (16.7b), the material in the space $0 < x < \Delta$ was taken to be of negligibly small viscosity, but we wish to adapt to the situation in Figure 16.2a where the two phases are more similar, and not separated by any weak film. In fact in Figure 16.2a, some rows of atoms are shown as continuous across the interface as a reminder that we wish to consider an ideally strong bond, and a stress profile more as in Figure 13.3. At such an interface, in the stiffer material the interface stress is *less* than its remote value, but in the less stiff material the interface stress is *greater* than its remote value. A diagram of the composition profiles on both sides must then be as in Figure 16.4, which makes an interesting comparison with Figure 16.2d.

Behavior affected by K^*: a stationary interface

At the start of this chapter, a three-part program was set out. The first two parts have been partly accomplished: ideas from Chapter 15 have been used in introducing eqn. (16.2), and ideas from Chapter 13 have been used in introducing Figure 16.4, but the two trains of thought do not yet interlock. To bring the two trains of thought together, we ask the following questions:

If two materials meeting at an interface are subjected simultaneously to change of temperature and to a constrictive strain, can these influences be adjusted in such a way that no change of phase occurs?

And, if this condition can be attained, what processes or changes will occur?

It will be proposed that the two influences can be balanced and that, even without the interface moving, compositions in the two phases will become or will be nonuniform; as elsewhere in nonequilibrium thermodynamics, a steady nonequilibrium condition can be maintained as long as a suitable flux of energy is driven through the system.

The first step is a minor change: in the first part of the chapter, the consequences were discussed of changing the *temperature* of a two-phase sample in a continuous way. Exactly similar changes could be produced by changing the overall *pressure* on the system. In many systems the effects are converse: raising the pressure and lowering the temperature have similar consequences. But this fact does not change the nature of the train of thought.

We consider then two phases, such as the phases g and h in Figure 16.2a, and let the phase with greater viscosity have the greater density. Both phases are, of course, of variable density; having compositions (A, B)X, the density of each varies with the abundance-ratio of A to B; but we suppose that the fundamental structure of the substrate X is less open in the high-viscosity phase, so that at any given A:B ratio it is the more dense phase. In other words, the molar volumes V^{AX} and V^{BX} are smaller in the phase where the viscosity N^* is greater.

Specifically, let phase g be the higher viscosity, lower-volume phase. Then, starting from an equilibrium condition, raising the overall pressure on the system will tend to make h convert to g. On the other hand, imposing a uniform constrictive strain rate will tend to make g convert to h, as in Chapter 13. If the two tendencies balance each other, an interesting condition will exist; it is this condition that we wish to explore.

To be specific, let the two phases coexist in equilibrium at some temperature T_e and uniform hydrostatic pressure P_e. Then let the temperature be kept at T_e throughout, let the magnitude of σ_{xx} be specified at any moment as $P_e + \delta p$, and let the imposed cylindrical constriction rate be e_0. Then in the equilibrium state (T_e, P_e), both e_0 and δp are zero; but we envisage a series of nonequilibrium stationary-interface states where e_0 and δp balance each other, neither being zero. What will such a state be like and how might it be reached?

If e_0 is not zero, then at points remote from the interface,

$$\sigma_{zz}(\text{remote})_g = \sigma_{xx} + 6N_g^* e_0 \qquad (16.8a)$$

$$\sigma_{zz}(\text{remote})_h = \sigma_{xx} + 6N_h^* e_0 \qquad (16.8b)$$

Suppose the profile of σ_{zz} is as shown in Figure 16.5; that is, suppose it continues to be of the same basic type as in Chapter 13. By itself, such a profile would drive self-diffusion of the ensemble (A, B)X across the interface without change of composition, just as in Chapter 13, with length scales B fixed by $(2N^*K^*)^{1/2}$. The material that travels is specifically wafers normal to z; the interface moves with respect to remote points because of nonuniform strain rates e_{xx} as shown in Figure 13.2d.

Let the processes just described run throughout a small interval of time δt; then if the velocity of the interface were v with respect to a homogeneously

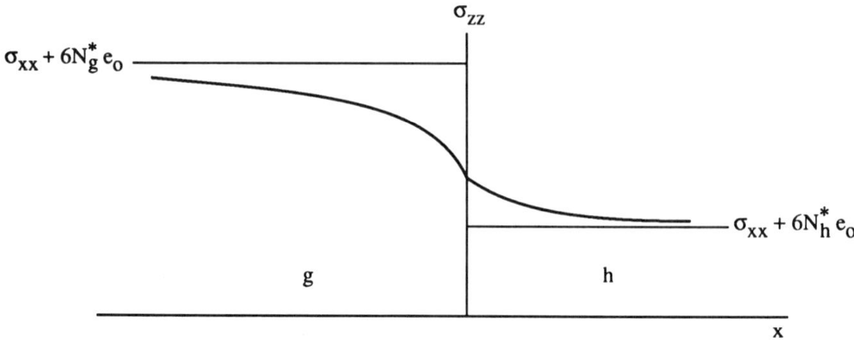

Figure 16.5 The stress profile at an interface where the constriction rate is e_0.

deforming grid that is pinned to the material at remote points, the displacement would be $v\,\delta t$. We wish to adjust δp so that during the same interval of time in absence of effect e_0, the displacement would be $v\,\delta t$ in the other direction; but this process affects a wafer or wafers normal to x. It appears that even a stationary interface is not inert: when δp and e_0 are in balance, a flux of wafers normal to z crosses the interface one way, and is matched by a more slow-moving flux of wafers normal to x moving the other way; see Figure 16.6. If the first flux occurred by itself, an excess of A atoms would be liberated at the interface and B atoms absorbed, but if the two fluxes run conjugately, no imbalance of A or B exists.

Regarding compositions at or around the interface, the first point to focus on is the interface itself. Here

$$\sigma_{xx} = P_e + \delta p$$

$$\sigma_{zz} = \sigma_{xx} + 6N_g^* e_0 - H_g = \sigma_{xx} + 6N_h^* e_0 + H_h$$

where H_g and H_h correspond to A_1 and A_2 in Chapter 13, and

$$\text{Mean stress} = (2\sigma_{zz} + \sigma_{xx})/3$$

which we designate as $P_{\text{interface}}$ or P_i. A diagram similar to Figure 16.2b represents equilibrium phase compositions at different pressures, all at temperature T_e. In such a diagram, compositions exist for the two phases g and h at the two pressures P_i and P_e; composition profiles in the condition e_0 and δp are then as in Figure 16.7, with a length-scale B_g fixed by $(2N_g^* K_g^*)^{1/2}$ and similarly for B_h. The profiles thus depend on the behavior of the ensembles $(A, B)X_g$ and $(A, B)X_h$ which gives N^* and K^* their values; but, being stationary profiles, their shapes do not depend on K^a, the coefficient reflecting the ease or rapidity with which species A and B exchange. (A small correction to this statement is linked to the slow flux of wafers normal to x, and is included in the next section.)

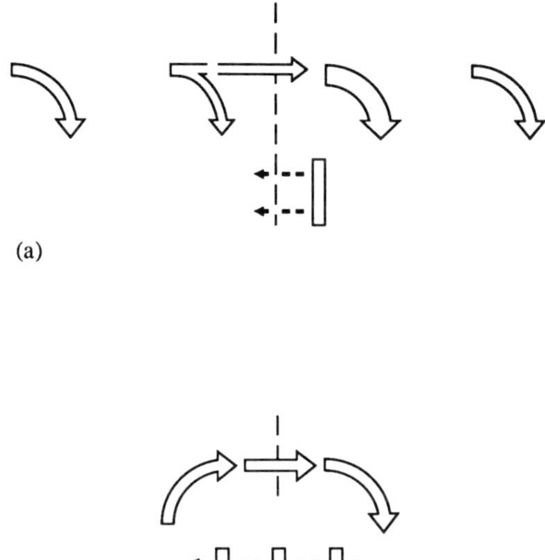

(a)

(b)

Figure 16.6 Material movements at an interface that remains stationary: (a) The total effect; (b) just the movements localized close to the interface, after subtracting movements related to the uniform overall strain. Each quarter-circle arrow represents a change of *shape*, i.e., a motion of wafers as in Figure 11.5b; all but one represent motions that lead to shortening along z and elongation along x.

The ideas just offered concerning a stationary interface are summarized in Figure 16.8. Diagram (a) shows a standard picture of the phase boundary at which phases g and h can coexist; the line is a line of true equilibrium states. Diagram (b) shows a plane of states for which the interface can be

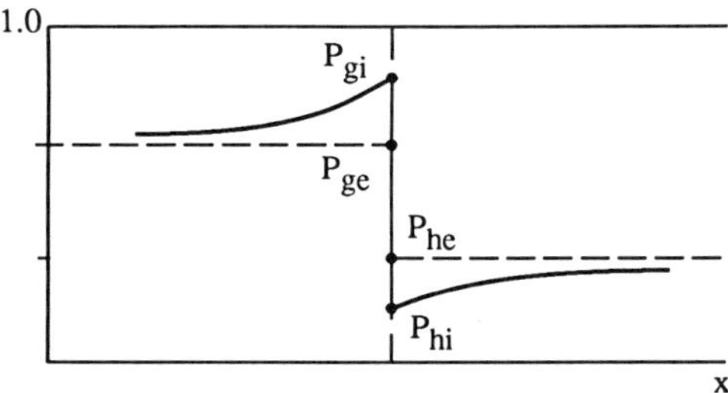

Figure 16.7 Composition profiles accompanying the stress state in Figure 16.5. The equilibrium compositions that were established throughout the phases at initial pressure P_e are marked P_{ge} and P_{he}; the compositions that develop at the interface at a new pressure P_i are marked P_{gi} and P_{hi}.

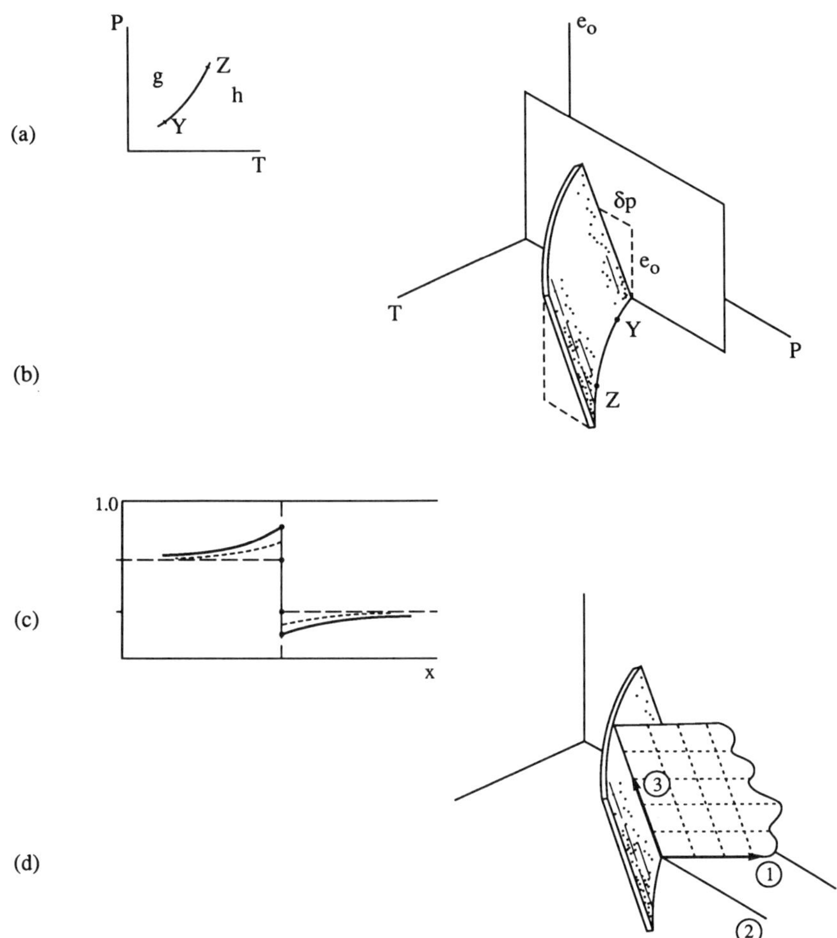

Figure 16.8 Conditions for a stationary interface. With no strain, the materials need to be in a condition on the line YZ in (a), the regular phase-boundary; but as shown in (b), for nonzero strain rates, there is a *plane* of possible conditions that give a stationary interface. At successively greater distances from the line YZ in this plane, composition profiles become successively more perturbed as in (c). For (d), see text.

stationary; at $e_0 = 0$, this plane contains the line in diagram (a), but where $e_0 \neq 0$, the surface is displaced; its slope in any plane $T = \text{constant}$ shows how e_0 and δp can increase while counteracting each other, giving stationary nonequilibrium states. The larger the conjugate values of e_0 and δp, the more vigorously the processes run that are illustrated conceptually in Figure 16.6, and the larger the variations in composition in the neighborhood of the interface, as shown in diagram (c).

The value of the diagram is that it permits the following general idea to be stated: at conditions represented by any point on line YZ, no processes run at all; at conditions represented by a point in the marked plane but not on line YZ, processes run that are controlled by the parameter K^*, with vigor proportional to the distance from YZ; but also, at conditions represented by points *not* in the marked plane, additional processes run that are controlled by the parameter K^a, the ease-of-exchange parameter, with vigor proportional to the distance from the stationary-state plane. In part (i) of this chapter, conditions were discussed that were not in the stationary-state plane, but there the constraint $e_0 = 0$ was used. In the next section, a reason will be given for changing that constraint; the simplest of conditions not in the stationary-state plane are found along some line such as line (1) in Figure 16.8d, rather than along line (2).

Behavior affected by K^a: a moving interface

The physical situation continues to be the one shown in Figure 16.2a, with two ideally coherent phases meeting. We are exploring the effects of changing the temperature and the stress state, and are seeking situations where quantities change with distance from the interface in simple exponential manner. If the phase relations are as in Figure 16.2b and an initial equilibrium state exists at pressure P_e and temperature T_e, then, as already discussed, if the temperature is changed at a rate that is slow but is increasing exponentially, profiles of composition and stress σ_{zz} can also take simple exponential forms $C_0 + ce^{kx}$ and $S_0 + se^{kx}$, if the magnitudes of the factors c and s satisfy eqn. (16.2a). This condition was established by considering a situation with no constrictive strain, $e_{zz} = 0$, but it also needs to be satisfied if e_{zz} is uniform along x according to

$$e_{zz} = e_0 = \frac{S_0}{6N^*} + [\quad]e^{kx} \tag{16.9}$$

Here [   ] denotes a set of five terms that total zero, comparable with those in eqn. (15.6a).

The need to consider nonzero magnitudes for e_0 is shown in Figure 16.9. An equilibrium situation is shown by full lines. It is imagined that by changing the temperature and so driving change of phase, nonequilibrium concentration profiles are created as shown by broken lines in the concentration diagram (a). For the concentration profile in phase g, a proportional stress profile $s_g e^{kx}$ can be conceived using eqn. (16.2a), and for the concentration profile in phase h, a stress profile $s_h e^{k'x}$ can be conceived similarly. Now there is no reason for the interface magnitudes s_g and s_h to be equal, so that if these stress profiles were superimposed directly on the equilibrium state "$\sigma_{zz} = P_e$ at all x," a mismatch or step in σ_{zz} would exist at the interface as in diagram (b). The stress profiles could not maintain their simple exponential form in presence of such a step, and *to eliminate the step*, it is

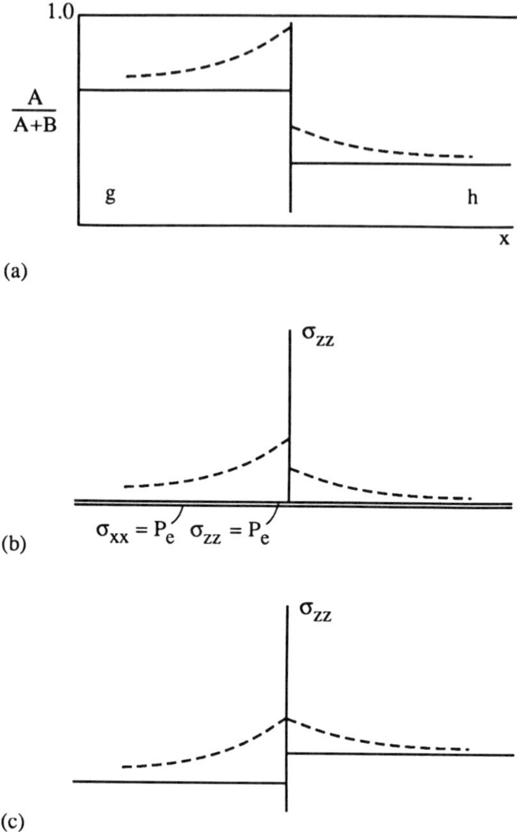

Figure 16.9 Conditions for simple proportionality of stress and composition as in eqn. (15.2). (a) Concentration profiles in equilibrium (full lines) and when the interface is moving (broken lines). (b) A stress profile compatible with the composition profile in phase g and a separate stress profile compatible with the composition profile in phase h; these two are not compatible with each other. (c) The stress profiles can be made compatible at the interface if their remote values are adjusted.

necessary to make S_0 in phase g different from S_0 in phase h. As long as the phases are of different viscosities, there must be some strain rate e_0 that, when imposed uniformly, produces the needed difference and yields an overall stress profile as shown in diagram (c).

Summing up, we envisage an interface initially at equilibrium. The temperature begins to change and its rate of change is slow but slowly increases exponentially. Exponential profiles of both composition and stress σ_{zz} can be maintained, whose width parameters stay constant while their gradients increase, the gradients staying in a fixed proportion. But to maintain this simple progressive history, one has to impose simultaneously a spatially uniform constrictive strain rate, whose magnitude has to increase with time as the temperature increases. All the above changes are taken to occur at some fixed magnitude of the interface normal stress σ_{xx}.

An alternative way of disturbing the equilibrium is to leave the temperature unchanged and instead to change σ_{xx} from its original value P_e by an amount δp that gradually increases. Again, to maintain simple behavior, the constrictive strain rate e_0 would have to increase; this time, its manner of increase would need to be properly related to the increase in δp. There must be some line such as line (1) in Figure 16.8d that shows how δp and e_0 need to change together. To make exponential profiles of concentration and stress σ_{zz} retain their widths and proportion to each other while increasing in gradient, one would have to drive e_0 and δp along path (1) at a gradually increasing rate.

Review

The chapter so far has comprised four headed parts. The first task is to note the difference between parts (iii) and (iv), i.e., between the stationary-interface condition and the condition just discussed. It is helpful to focus on the length-scale and the *primary* diffusion parameter; in part (iv) these are $1/k$ and K^a, whereas in part (iii) they are B and K^* (with $B^2 = 2N^*K^*$). In part (iv), k has no fixed value: as discussed in part (i), if conditions move away from equilibrium slowly, k is small and profiles have small gradients, whereas if conditions change faster, k is larger. By contrast in part (iii), the main effects are all on length scales B_g and B_h which are almost-invariant fundamental properties of the two phases (almost invariant as long as one mechanism continues as the mechanism for change of shape and diffusion; when the dominant mechanism changes, B changes).

In part (iii), the main processes are continuum processes—deformation and self-diffusion of the substrate X. In part (iv), the main processes are particle processes—exchange of A for B in response to changing temperature or stress is the stimulus for all that goes on.

Even in part (iv) where K^a is dominant, K^* is not wholly eliminated; it plays a part in fixing the $s{:}c$ ratio at which all the processes run. But the rate at which they all run is directly and primarily controlled by K^a. Conversely, in part (iii) where K^* is dominant, K^a is not wholly eliminated; the stationary concentration profile has to stay stationary despite the circulation of material through it, as in Figure 16.6. If the rate of that circulation were to increase, the concentration profile would adjust itself slightly, and the amount of adjustment needed depends on K^a; the parameter K^* is the direct and primary control but K^a has this minor role.

Another way of seeing the relation of parts (iii) and (iv) is through Figure 16.10. The dotted lines are profiles of σ_{zz} copied from Figure 16.9b and showing a step at the interface. Such a condition could not exist: as discussed in Chapter 13 and part (iii) of this chapter, it would be smoothed off, but the effect of the smoothing would be negligible at distances greater

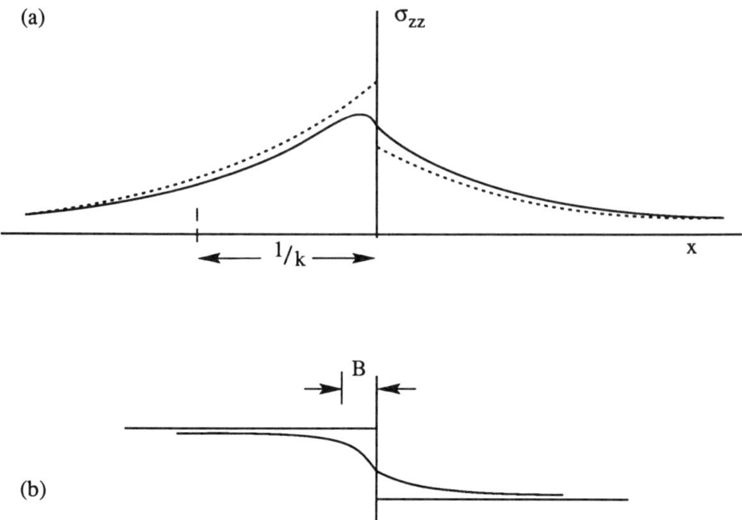

Figure 16.10 (a) The incompatible stress profiles from Figure 16.9 are shown dotted and a possible actual stress profile is shown by a full line. The length scale for the dotted profile is $1/k_g$ or $1/k_h$, whereas the length scale for the difference between dotted and full lines is B. (b) The difference between dotted and full lines plotted separately; compare with Figure 13.3b.

than $3B$ from the interface. The full line shows a possible result. The basic dotted-line curves are from part (iv) but the smoothing of the step is from part (iii), for which diagram (b) is a reminder. By using an appropriate rate e_0 as in part (iv), the step is eliminated and no effects on length scale B remain. By contrast, another appropriate rate e_0 as in part (iii) suppresses movement of the interface and the dotted-line effects, so that *only* effects on length scale B remain.

The two length scales help us to organize our thoughts in two ways. The first has just been noticed: length scale B reminds us of the material as an ensemble, with properties K^* and N^* and behavior as in Chapter 13; by contrast, length scale $1/k$ reminds us of the material as a matrix within which exchange processes occur, with dominant property K^a and behavior as in regular interdiffusion (with no deformation effects superimposed). The second use of the two length scales is to help us grapple with the full line in Figure 16.10a, the compressive stress profile in a nonspecial condition. This is shown in yet more general form in Figure 16.11. The suite of profiles is in some sense the end of our quest, or the end of stage 1 and the start of stage 2: in constructing it, we use the ideas that the material behaves partly as a continuum and partly as atoms, and that it is driven by both physical stress and chemical inhomogeneity; recognition of the one profile shows, looking backward, that these somewhat conflicting ideas can be combined. But also, looking forward, we see mathematical difficulties looming: at key

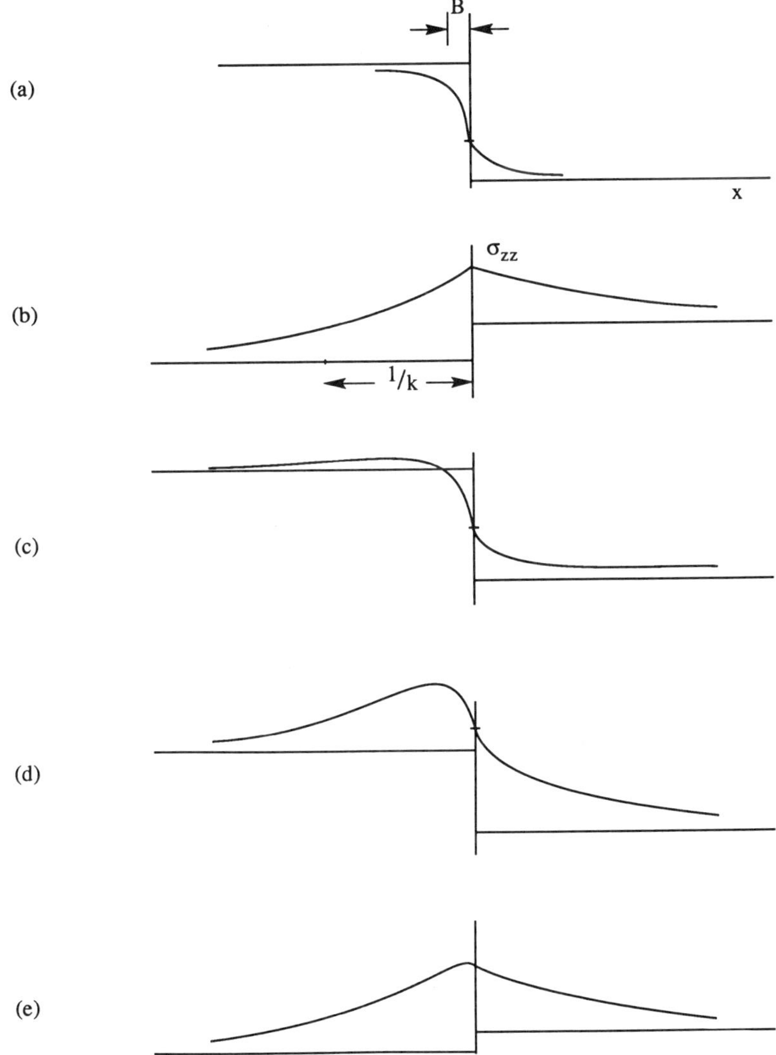

Figure 16.11 Possible stress profiles that combine effects on two length scales. (a) A stress profile with length scale B, as in Figure 16.5. (b) A stress profile with length scale $1/k$, as in Figure 16.9c. (c) A sum of curve (a) plus a small proportion of curve (b). (d) A sum of curve (a) and curve (b) in equal proportions. (e) A sum of curve (b) plus a small proportion of curve (a). In the figure, curves (c) through (e) are created strictly by addition. In a real material, the stress profile would arise from interplay of the mechanical and chemical influences and would not be related to the special-case profiles by simple arithmetic. But the range of possible real profiles is expected to have the same general properties as the range of curves (c) through (e).

points in Chapters 14 through 16, we were able to discard a factor $\sin jx$ or e^{kx} from an equation because the same factor appeared in all terms, and thus reach a simple result, Figure 16.11a or 16.11b. But Figures 16.10a and 16.11c–e involve two length scales simultaneously, and it cannot be expected that in future the length-scale terms will disappear so conveniently. The suite of profiles being thus a milestone or staging point, we break off and devote a chapter to reviewing where we have been and in what direction we might next wish to head.

APPENDIX 16A: MAGNITUDES OF e_0 AND δp

The purpose of this appendix is to inspect Figure 16.8d and consider what quantities need to be known in order to construct such a diagram in numerical form. In considering the quantities, we try to estimate an order of magnitude for each, but use the coarsest and simplest assumptions; the objective is simply a first sketch of a quantitative approach.

Stationary interface

The condition of interest is illustrated in Figure 16.6. Here the open arrows represent effects discussed in Chapter 13; these by themselves would lead to change of phase or migration of the interface through the material. But the solid arrows show a compensating effect, so that the set of processes all operating together keep the interface stationary. We consider the open arrows and then the solid arrows in turn.

The stress profiles that drive the open-arrow processes are shown in Figure 16.5 and are described by eqns. (13.5) and (13.8). To see how fast the interface would move, we estimate the material flux across it as follows.

The stress gradient $\partial\sigma_{zz}/\partial x$ adjacent to the interface in phase 1 is A_1/B_1 or

$$\frac{K_2}{B_1 K_2 + B_2 K_1}\, 6e_0(N_1 - N_2)$$

With cylindrical symmetry and σ_{xx} uniform along x, the material flux is then

$$\frac{2K_1}{3}\frac{K_2}{B_1 K_2 + B_2 K_1}\, 6e_0(N_1 - N_2) \qquad \text{or} \qquad \frac{4K_1 K_2}{B_1 K_2 + B_2 K_1}\, e_0(N_1 - N_2)$$

The material flux adjacent to the interface in phase 2 is the same (as it needs to be to conserve material). The units are (m^3/m^2-sec) or (m/sec): if x m^3 of material cross 1 m^2 of interface per second, the interface moves at x m/sec with respect to a homogeneously deforming grid that is pinned to the material at remote points.

To estimate orders of magnitude, assume the simple condition $B_1 = B_2$; let their common value be B. Then the interface velocity becomes

$2Be_0(N_1 - N_2)/(N_1 + N_2)$, and if N_2 is appreciably smaller than N_1, for example $N_1/10$, the velocity is adequately represented just by $2Be_0$ (m/sec).

For examples of numerical values, let us replace e_0 by $(\sigma_{zz} - \sigma_{xx})/6N$. Let $\sigma_{zz} - \sigma_{xx}$ be 100 MPa or 10^8 Pa: then, using values of N and B for glass at 1100 K from page 120, we estimate a velocity of 1 µm/sec; or, using values for olivine at 1700 K, we estimate a much smaller velocity 15×10^{-10} µm/sec.

The question then arises: If two phases are in equilibrium under hydrostatic conditions P_e and T_e, by how much must the pressure be made to differ from P_e in order to make the phase boundary move at 15×10^{-10} µm/sec? Somewhat relevant observations are reported by Vaughan et al. (1984): their run 92 at 1563 K lasted 74 hours, produced migration distances of the order of 10 to 50 µm and probably contained stresses differing from P_e by a few hundred megapascals. The apparent interface velocity is about 10^{-4} µm/sec. We conclude that probably a δp magnitude much less than 1 MPa would drive the interface at the required velocity. In other words, the solid-arrow processes in Figure 16.6 require magnitudes for δp that are *much* smaller than the magnitudes of $\sigma_{zz} - \sigma_{xx}$ that drive the open-arrow processes.

The preceding remarks are more to give a numerical example that works in principle than to reach firm conclusions. The data set contains several large uncertainties. As noted before, data sets are few: there are many materials whose creep has been measured, but very few where it has been measured at sufficiently low stress for the behavior to be linear and where any measured rate of phase-boundary movement has been linked to pressure.

Moving interface with stresses equalized

The relevant illustration is Figure 16.9 and the feature of interest is the step or mismatch in stress magnitude σ_{zz} at the interface in Figure 16.9b. We wish to estimate a constriction rate that, if imposed uniformly, would eliminate this step. Again the objective is to make just a preliminary approach to orders of magnitude.

The two stress magnitudes that fail to match have been designated s_g and s_h. Either of these is given by an equation of type (16.2). As already discussed, this equation is not reliable when the denominator approaches zero; but if we go to conditions where the denominator is not close to zero (where the factor $1/N^*$ is the dominant factor), the equation resembles eqn. (15.2). This equation is illustrated numerically on page 146, where it appears that if $1/N^*$ dominates the denominator, s must be 100 MPa or less. If both s_g and s_h are of the order of 100 MPa, their difference must be somewhat less; strictly for purposes of illustration, let us suppose that $s_h = 0.7s_g$ and the difference $= 0.3s_g$—for example, 30 MPa or less.

If two phases with viscosities N_g and N_h are constricted at the same radial strain rate e_0, the driving stress differences in the two phases are $6e_0N_g$ and $6e_0N_h$. If the stress states have a common stress σ_{xx}, the two magnitudes of σ_{zz} will differ by $6e_0(N_g - N_h)$. If N_h is, for example, $N_g/10$, the order of magnitude of the difference is simply $6e_0 N_g$. Then the constriction rate e_0 that would create a correction of 30 MPa would be (30 MPa)/$6N_g$ or in general $(0.3s_g)/6N_g$ or $s_g/20N_g$. The factor 0.3 is arbitrary but its magnitude is of the right order.

Representation in a diagram

The quantities discussed above can be partly represented in diagram form; see Figure 16.12. The figure is a cross-section through Figure 16.8d in a plane of constant temperature. The equilibrium phase boundary pierces such a plane at a single equilibrium point, shown with the label (P_e, T_e) in Figure 16.12. Let scales be chosen for the axes and let a line be drawn through (P_e, T_e) with gradient equal to $1/6N_g$. Then, for example, if σ_{xx} is maintained equal to P_e, any difference $\sigma_{zz} - \sigma_{xx}$ can be plotted along the stress axis and the vertical height to the line shows the corresponding constriction strain rate.

The moving-interface, matched-stress condition is illustrated by the dotted rectangle. We suppose that equilibrium is established and then conditions are slowly changed; the interface begins to move and interdiffusion begins to run, and *some* stress magnitude s_g develops in phase g. Then from the discussion just given, we identify point Q, whose distance above the stress axis is 0.3 times the height of the rectangle; point Q fixes the direction of line (1).

The stationary-interface condition is illustrated by the broken-line rectangle. Here we begin by imposing a strain rate E; then the length of the rectangle shows the stress difference $\sigma_{zz} - \sigma_{xx}$ that develops at remote points

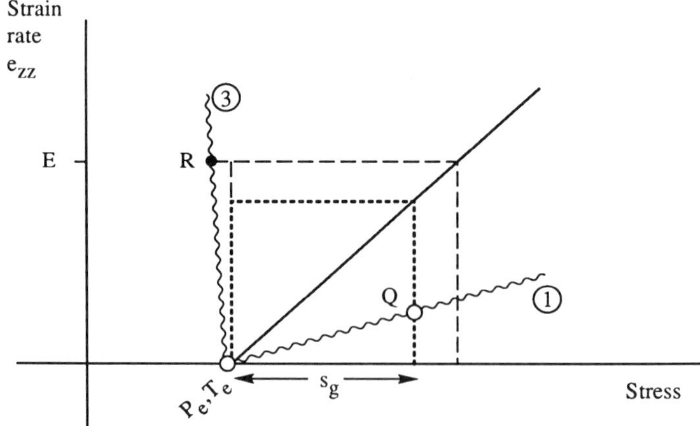

Figure 16.12 Relations in a plane of constant temperature from Figure 16.8. Lines (1) and (3) correspond with the same lines in Figure 16.8d.

in phase g. From E, we can also calculate an interface velocity $2B_g E$, and show the needed countervailing magnitude of δp as a line-segment ending at point R. Point R fixes the direction of line 3. When olivine is used as example, the deviation of line 3 from vertical is wholly negligible, and one might conjecture that this is likely to be the common situation.

17

SUMMARY

The purpose of this chapter is to consolidate. No new ideas are introduced; instead we try to sort the main thread from the side issues, and the parts that are reasonably clear and firm from the parts that are still fuzzy. The core of the chapter is a set of seventeen statements, seventeen vertebrae that form the backbone of the book, but there are also a preface and a postscript. The preface provides the setting for the seventeen-part core and the postscript takes up the question of where to go next.

Preface

The purpose of the book was given at the start of Chapter 1. Even at that early point, a stressed cylinder was used as an example. The purpose is to make headway with the question: if a state of chemical equilibrium exists under hydrostatic stress and is disturbed by making the stress nonhydrostatic, what processes begin to run, and what quantitative relations should we expect to be followed? Before the seventeen-part "answer" it is to be noted that there are two alternative ways of dividing the subject matter into two parts.

The division scheme is displayed in Figure 17.1a and separates eight types of change. (A somewhat similar diagram on page 111, distinguished eight circumstances in which change might be observed—a different system of divisions that is of no use here.) Of the eight boxes set up, four have been discussed, as shown in Figure 17.1b. The two ways of dividing this four-box group are by a horizontal cut or by a vertical cut that separates stars from superscript a's. (A vertical cut separating the N-box from the rest is of no help; it would be contrary to our theme.)

The horizontal cut separates stress-driven effects below from composition-driven effects above. It is in fact the traditional division between mechanics and chemistry; enormous amounts of science fall clearly above the cut or clearly below it and cause no confusion at all. This cut was used as a guide in the early chapters, especially in the flow diagram or organization chart, Figure 8.1.

By contrast, the second cut appeared as late as Chapter 15, but deserves emphasis; it is at least as instructive and helpful as the first, and perhaps more helpful. It distinguishes behavior where two components behave in parallel or go along with each other (star boxes) from behavior where they act in opposite ways (boxes with superscript a).

174

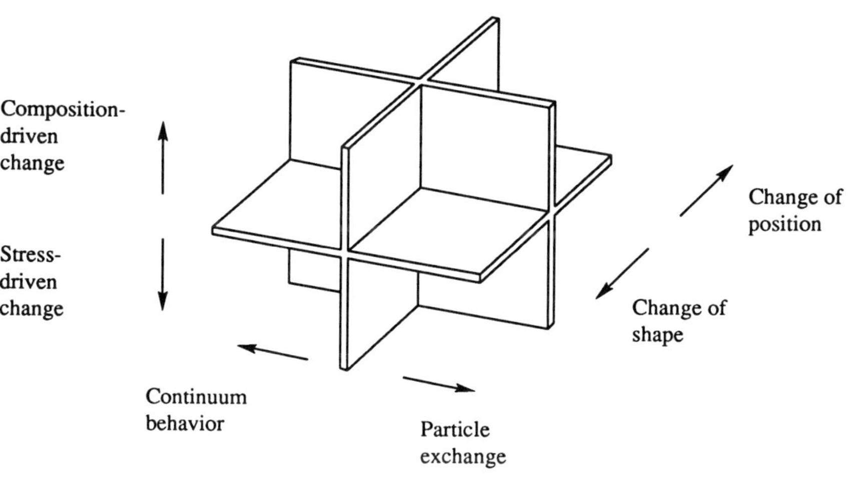

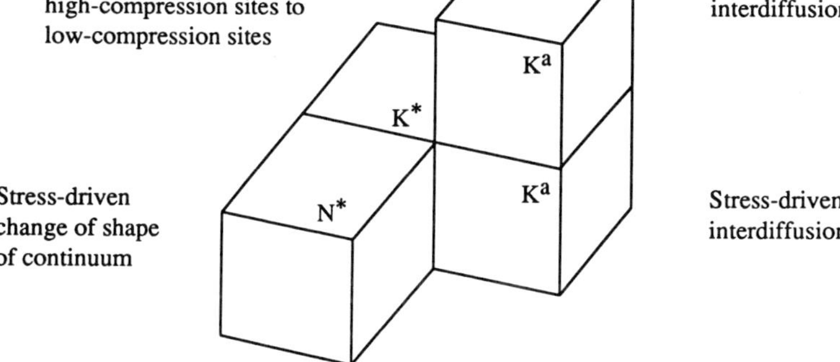

Figure 17.1 An overview. (a) A framework containing eight distinguishable types of behavior. (b) Locations within the framework of the four types of behavior discussed so far, with their controlling parameters indicated. *Continued on p. 176.*

The distinction just made is very clear in Chapter 15 where compounds of type (A, B)X are discussed. In this context, A and B go along with each other in a very specific way: the total number of A plus B in any designated volume-element equals the number of X, and when the element deforms or changes content by self-diffusion, the abundance-ratio of A to B can be treated as staying constant. We do not by any means believe that it does stay constant but if, during deformation, the ratio changes, the change can

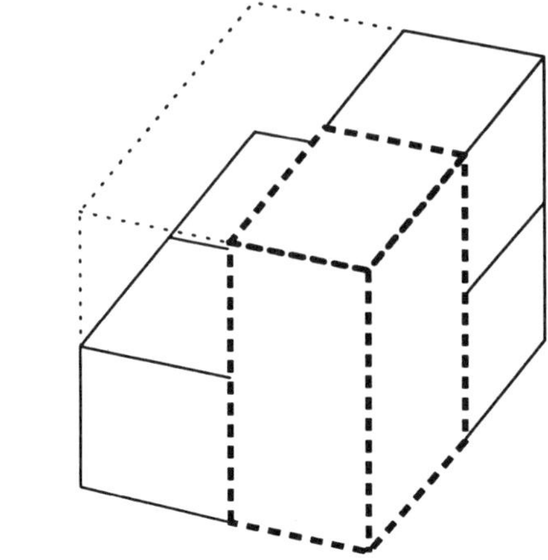

(c)

Figure 17.1 *continued* An overview. (c) Behavior in the broken-line box could occur in certain circumstances, whereas behavior in the dotted-line box seems impossible, a contradiction in terms.

be considered a consequence of superimposing the *other* behavior, interchange of A and B within a set of sites of fixed number.

As an example, consider stress-driven self-diffusion from a high-compression site to a low-compression site. In theory each of the species A, B, and X has a different partial molar volume, and hence is subject to a different potential difference between the two sites, ΔPV^A, ΔPV^B, or ΔPV^X if there is a pressure difference between the two sites of ΔP. But in fact, because of the stoichiometry constraint or charge-balance requirement, what moves by self-diffusion is a composite unit $(A_cB_{1-c})X$ with effective volume between $(V^A + V^X)$ and $(V^B + V^X)$: it is a weighted mean of V^A and V^B that helps to fix the flux of this joint self-diffusion regardless of whether V^A and V^B are close or noticeably different. By contrast, the difference between V^A and V^B fixes the vigor of their interdiffusion (that is, the extent to which they move with respect to each other) and this effect is regardless of whether the values V^A and V^B are both large or both small. To know the magnitude of V^A does not enable us to predict either of these effects; even knowing the magnitudes of both V^A and V^B does not enable us to predict them *until* we transform the information into a mean value and a difference.

The distinction now being emphasized between a-terms and star-terms was shown diagrammatically in Figure 15.3b with black and white arrows. It is a distinction easy to make in (A, B)X compounds and less easy to make in simple binary materials. One purpose of this review chapter is to identify aspects that are not yet fully worked out, that are currently fuzzy, and binary materials such as alloys are one such aspect. Clearly, to some extent binary alloy behavior can be described as joint behavior with some interdiffusive or exchange behavior superimposed on it, but in this context it is not so clear whether the exchange should be thought of as atom-for-atom or volume-for-volume or mass-for-mass; perhaps for one purpose one basis would be more useful and for a second purpose another basis might be more useful. This uncertainty reappears on page 201; for this review, we simply note it as a topic that is not yet clear-cut.

With Figure 17.1b in view we can restate the purpose of the book: (1) the purpose is to explore ways of predicting the behavior of materials where movement is driven simultaneously by nonhydrostatic stress and by chemical inhomogeneity *or* (2) the purpose is to explore ways of predicting the behavior of materials where the response of the material as a chemically invariant continuum has extra behavior superimposed on it, in which individual components exchange sites or interdiffuse.

The box "stress-driven interdiffusion" is the one on which attention centers, which can be treated in either of two ways. We envisage here a crystalline material that, despite its crystallinity, can be made to creep (like a glacier): if it is squeezed in a nonuniform way, its composition changes—it deforms as a continuum and changes composition by particle exchange; to the one act of squeezing, the material responds mechanically and chemically.

Because of this box, the book has a characteristic: it mixes discussion of a continuum with discussion of the behavior of atoms. This is not a defect, it is not due to carelessness or a failure to be well organized. Continuum concepts and atomic concepts are combined because that is the purpose. The core of the undertaking is to construct equations that combine terms based on continuum concepts with terms based on atomic concepts, and still have the equations yield believable (ultimately testable) predictions about real material behavior.

Besides ending the comments on Figure 17.1, the previous sentence serves as the first of the vertebrae to which we now turn.

Core

• The objective is to construct equations that combine terms based on continuum concepts with terms based on atomic concepts, and that yield predictions about real material behavior.

• The deformation of a continuum is described by imagining small planar elements. Among the things one has to imagine is a pair of parallel planar elements whose separation changes. If the continuum is of constant density, such a change of separation can be described in terms of wafers, and in terms of the limit of a series where the wafers become successively smaller in size (Figures 1.1 and 11.4).

• Materials of constant chemistry and constant density can self-diffuse. If an ideal continuum is to be used for building equations and is to represent real materials, it must show self-diffusion behavior. Both self-diffusion through space and change of shape of a continuum "at a point" can be described as the behavior of wafers. In self-diffusion the wafers do not change orientation, whereas in change of shape the wafers do change orientation (Figures 1.2, 11.5, and 11.6).

• Migrating material has no sense of orientation. Material travels along a path because of a gradient along the path, and the relation of flux to gradient must be the same whether the path is curved or straight.

• In a nonhydrostatic stress field that is homogeneous through space, a material's response is *as if* wafers travel on curved paths that are specifically quadrants. The gradient that drives the wafers is the gradient in the stress component normal to the wafer. The agent driving motion is clearly not a gradient in the mean stress because there is no gradient in mean stress along such a path.

• From the preceding point, two conclusions are drawn. First, if material self-diffuses from a source-region to a sink-region because the stress field is not homogeneous through space, this motion also is driven by gradients in normal-stress components rather than a gradient in mean stress.

Second, there is an identifiable length, a number of nanometers, here designated L_0, such that the material's resistance to change of shape "at a point" is the same as its resistance to spatial self-diffusion through a distance L_0. The material's self-diffusive and viscous behavior is *as if* it changed shape by migration of wafers along quadrant paths that are specifically L_0 nanometers in length.

We should distinguish the real from the imaginary. In the preceding sentence the phrase "the material's behavior" refers to real behavior, as might be observed in an experiment. To predict behavior, we imagine an ideal continuum subject to laws expressed in equations; it is in this continuum that material is imagined to migrate along quadrants that have finite length L_0 while being infinitely thin, infinitely numerous, ideally set up to cross one another without communicating, etc. The function of the continuum is to obey equations exactly and deliver predictions.

The five preceding points are summed up in eqn. (12.7): for some direction **n** in a material

$$\text{(strain rate)} \quad e_{nn} = \frac{K}{3}\left[\frac{\partial^2}{\partial(R\alpha_1)^2} + \frac{\partial^2}{\partial(R\alpha_2)^2} + \frac{\partial^2}{\partial x_l^2} + \frac{\partial^2}{\partial x_m^2} + f\frac{\partial^2}{\partial x_n^2}\right]\sigma_{nn}$$

$$(12.7)$$

Here K is a self-diffusion coefficient, $R = 2L_0/\pi$, and f is one of the features currently still fuzzy. It lies between 0 and 1; we defer elucidating f because, in all the situations now under review, the term containing f is taken as negligibly small anyway.

The simplest applications of eqn. (12.7) are to situations with cylindrical symmetry. Let x_m be the direction of the symmetry axis and let the situation be uniform in all respects in transverse directions x_l and x_n; then $\partial^2\sigma/\partial x_l^2 = \partial^2\sigma/\partial x_n^2 = 0$. Also, if α_1 is taken to lie in the mn plane, then α_2 lies in the ln plane and $\partial^2\sigma/\partial\alpha_2^2 = 0$. There remain just two contributions to the transverse strain rate e_{nn}, a contribution from constrictive change of shape and a contribution from self-diffusion along the axis.

Two examples follow. First, let the transverse compression σ_{nn} vary harmonically along m: then the constrictive strain rate also varies harmonically. Where σ_{nn} is a maximum, $\partial^2\sigma_{nn}/\partial x_m$ is also a maximum and the two contributions to e_{nn} reinforce each other; both vary harmonically and are in phase. Second, let the sample consist of two polymorphs of different viscosities as in Chapter 13: then a *uniform* constrictive strain rate leads to nonuniform transverse compression. Profiles of σ_{nn} are exponential in such a way that where σ_{nn} is less than its remote value, $\partial^2\sigma_{nn}/\partial x_m^2$ is greater than its remote value. The two contributions to e_{nn} vary antipathetically and can add up to a constant sum at all points along the cylindroid axis x_m.

All ideas to this point come from continuum mechanics; ideas about chemical potential and change of chemistry follow.

- *Chemical potential.* Let a thermodynamic system in equilibrium contain, among other components, a mass m_i of component i. Let the enthalpy or Gibbs free energy of the system be G. Let a reversible change occur in which m_i increases by a small amount δm_i with an accompanying increase in G of δG. Let the ratio $\delta G/\delta m_i$ be followed to the limit as δm_i goes to zero, and let the limit be designated μ_i, the chemical potential of component i in that system.

The thermodynamic system in view occupies a certain volume of space and has a bounding surface but, in the ideal state of equilibrium, it makes no difference where the new mass δm_i is lodged or across which part of the bounding surface it passes to enter the system; μ_i is not affected by any such detail.

- *Nonequilibrium systems.* Consider a system that is not in equilibrium because conditions in one part are different from conditions in another part. Immediately we notice two facts: first, there must be some kinds of gradient through space from one part to the other; second, the possibility appears of finding an equilibrium state that closely resembles the state in one part of the system in view, and a *different* equilibrium state that closely resembles the state in another part. In each of the two equilibrium states thus identified, a component i has a potential μ_i. Let these potentials be μ^A in part A of the system and μ^B in part B: then if the separation of the parts is a distance d, one can define a potential gradient as $(\mu^A - \mu^B)/d$.

Such a gradient is slightly paradoxical: to define μ^A we have to imagine an equilibrium system, that is, a system with no gradients; yet we use the magnitude of μ^A to establish the presence of a gradient. The key is in the words "closely resembles" and Figure 4.1 illustrates how the ideas link up.

- *Nonuniform pressure.* As an example of nonequilibrium, we might consider a system where one part is close to hydrostatic at pressure P^A and another part is close to hydrostatic at pressure P^B. Part A would have some kind of limits to its extent, some kind of bounding surface, and if a mass δm_i of component i were inserted across the boundary, the change δG would approximate $\delta m_i \mu_i^A$. By contrast if the mass δm_i were inserted across the boundary of part B, the change δG would approximate $\delta m_i \mu_i^B$.

- *Nonhydrostatic systems.* A material whose stress state is homogeneous through space but nonhydrostatic resembles the above. Let the maximum and minimum principal stresses be σ^A and σ^B and let us imagine two corresponding hydrostatic states, one at pressure P^A and the other at pressure P^B, where $P^A = \sigma^A$ and $P^B = \sigma^B$. Then the process of introducing a mass δm_i across a part of the system's bounding surface with normal stress σ^A resembles introducing it into a hydrostatic system at P^A, while the process of introducing the mass δm_i across a bounding surface at σ^B resembles introducing it into a hydrostatic system at P^B. Again we get two distinguishable potentials μ_i^A and μ_i^B.

The train of thought is not restricted to principal planes of the stress state but applies to any plane. At some point in the material, for any direction through the point there is a normal-stress component σ_n with magnitude between σ^A and σ^B, an associated imaginary hydrostatic pressure P^n and an associated equilibrium potential μ_i^n. The associated equilibrium potential is a direction-dependent scalar with an infinite number of magnitudes at a point, just like the normal-stress component of a stress state. The two quantities are linked by the factor V_i, the volume of one unit of component i (1 kg or 1 kg-mol or other unit):

$$\frac{\partial \mu_i^n}{\partial \sigma^n} = \frac{\partial \mu_i^n}{\partial P^n} = V_i$$

For a material of only one component, this proportionality yields eqn. (12.8):

$$\text{(Strain rate)} \quad e_{nn} = \frac{K}{3V}\left[\frac{\partial^2}{\partial (R\alpha_1)^2} + \frac{\partial^2}{\partial (R\alpha_2)^2} + \frac{\partial^2}{\partial x_i^2} + \frac{\partial^2}{\partial x_m^2} + f\frac{\partial^2}{\partial x_n^2}\right]\mu_n$$

For examples of the use of eqn. (12.8), the same two situations can be examined that were used to illustrate eqn. (12.7). Again the five terms can be reduced to two terms by assuming uniformity and symmetry; one term in e_{nn}^{total} is a contribution from constrictive change of shape and the other is from self-diffusion along the axial direction x_m.

• At this point, we have adapted the concept of chemical potential so as to apply it in situations where the stress field is nonhydrostatic as well as nonuniform through space. We have related a material's strain rate or change-of-dimension behavior to its chemical-potential field. But so far we have discussed only problems of continuum mechanics type—the same problems that we were able to discuss effectively in terms of stress. We have used chemical potential in describing continuum-mechanics behavior, the left-hand two boxes in Figure 17.1; but we have not yet used chemical potential in describing change of concentration of an atomic species, which is of course one of the concept's most natural and powerful uses.

• The simplest material that can change composition is a binary mixture of two atomic species; an example would be a copper–tin alloy. Let a mixture of this type contain atomic species A and B: then for the chemical potential of species A, $\partial\mu^A/\partial C^A = RT/C^A$, where C^A is the mole fraction or number ratio of atoms of A to total number of atoms in a sample, $n^A/(n^A + n^B)$, and $\partial\mu^A/\partial P = V^A$, the partial molar volume of species A. If stress is nonhydrostatic, the associated equilibrium potential of A for direction $\mathbf{n}$ follows a similar relation $\partial\mu_n^A/\partial\sigma_{nn} = V^A$.

The objective is to discover how the chemical behavior of such a material is affected by a condition of nonhydrostatic stress.

• As before, a situation can be chosen for study so that in the strain-rate equation (12.8), the five terms reduce to two terms—an orientation term $\partial^2/\partial\alpha^2$ and a spatial position term $\partial^2/\partial x^2$. Suppose μ^A varies with composition and stress as just noted in the previous paragraph: then the strain rate of component A gains in principle four contributions, from change of stress with position, stress with orientation, composition with position, and composition with orientation. But the last is disregarded, so that three contributions to strain rate remain.

The second component B furnishes three more contributions. If a condition is imposed that the sample's total strain rate along direction $\mathbf{n}$ be some fixed value such as zero, a relation among six terms results, the six terms being

> stress-driven change of shape of A at the site,
> stress-driven diffusion of A to or from the site,
> composition-driven diffusion of A to or from the site,

and similarly for B. In the simplest of illustrative situations, all stress-driven changes are governed by a single stress difference, and all composition-driven changes are governed by a single composition difference. Then the six-term relation becomes a relation between these two, and achieves the purpose given at the outset: if an initial equilibrium state of a binary material is disturbed by making the composition nonuniform, the relation shows what nonhydrostatic stress state would satisfy the boundary conditions, *or* if equilibrium is disturbed by making the stress nonhydrostatic, the relation shows what nonuniformity of composition would satisfy the boundary conditions.

• Although a path to results of the desired type has been outlined, there remains room for variation and exercise of choice. The six terms in view can be indicated by

$$\frac{s}{N^{A}}, \qquad K^{A}s, \qquad E^{A}c, \qquad \frac{s}{N^{B}}, \qquad K^{B}s, \qquad E^{B}c$$

Here s or c is the driving agent, either stress difference or composition difference, N^{A} and N^{B} are viscosities controlling rate of change of shape, K^{A} and K^{B} are mobilities controlling stress-driven migration, and E^{A} and E^{B} are mobilities controlling composition-driven migration. In the terms s/N^{A} and s/N^{B} each component is treated as a continuum, whereas in the terms $E^{A}c$ and $E^{B}c$ each component is treated as a set of particles (specifically 6×10^{26} independent particles per kg-mol). There is room for exercise of judgment in deciding how to modify the primitive ideas given so as to reduce this type of clash. We seek equations where the concepts that underlie one term do not conflict too strongly with the concepts that underlie the next term, and there is no clear-cut best modification to make; best procedure for alloys may be different from best procedure for polymers and different again from best procedure for ceramics.

• *Materials of formula* (A, B)X. The materials are taken to be ideally stoichiometric, with the total number of atoms of species A and B equal to the number of atoms of X in any volume. Then the material can be viewed as a simple binary combination of units AX and units BX. On the other hand, it can alternatively be viewed as predominantly a substrate BX, with a small volume-fraction of an additive, as in Figure 15.1. In the second view, both parts contribute to the volume occupied, but only the substrate has continuum properties and only the additive has particle properties (being affected by gradients in concentration, for example). This view enables us to assemble equations, for example for total rate of change of volume of a portion of the ensemble, that contain terms based on continuum concepts plus terms based on particle concepts, yet we avoid treating any single material in both manners simultaneously (and that is a thing we want to avoid, because it is self-contradictory).

• *Materials of formula* (A, B)X *at an interface.* As an example, the procedure outlined enables processes close to an interface to be analyzed in simple terms. Suppose that two phases are initially in equilibrium across a planar interface but that conditions are changed so that the equilibrium is upset. Processes begin to run and gradients develop that are strong close to the interface and weaker farther away. For the type of material in view, continuum effects would by themselves diminish outward in a simple way on a short length scale, and particle-exchange effects would by themselves diminish outward on a longer length scale. If either effect were present alone, the transverse compressive stress could also change in a simple concordant way. But if both effects are present simultaneously, the one profile of transverse stress cannot be concordant with either, and all simple relationships are lost.

● The approach using substrate and additive permits progress as follows: it enables us to discover a series of conditions along which we can move away from equilibrium while initiating processes on only one of the two conflicting length scales—and also a second series of conditions along which we can move while initiating processes on only the other. These two series form the beginnings of an answer to the question: what processes begin to run when equilibrium at the interface is nonhydrostatically disturbed? They are early steps toward a more complete analysis.

Postscript

The sequence of topics up to this point has been more or less linear; at least, it has been represented as linear in the chapter-sequence diagrams, on pages 54 and 111. From this point on, however, there are definitely two divergent tracks. One leads on from a simple planar interface to behavior around an inclusion—first a single inclusion in an extensive matrix and then some nonquantitative remarks about an aggregate of grains. Along this track, the intention is, if possible, to make definite statements; that is, as elsewhere, to find conditions simple enough to permit definite statements to be made. By contrast, along the other track the intention is to express doubt: not to progress farther but to go back over previous ground and draw attention to points that are not clear or not firm, such as the factor f in the five-part strain-rate equations (12.7) and (12.8). There are indeed much more substantial points of doubt than this one.

Because the arrangement of topics is less linear than before, readers will probably skip and be selective. The whole book is tentative and exploratory and, especially from this point on, what is presented is only sketches. It is supposed that readers will not only skip and be selective but will treat the text as transient, to be replaced by revisions and supplements of their own.

IV
EXTENSIONS

18

CYLINDRICAL INCLUSIONS

The purpose of this chapter is to extend the ideas in Chapter 16 to the situation where one material occurs as an inclusion in the other.

In Chapters 13 through 15, most of the discussion centered on conditions that varied along one direction, x, but not along orthogonal directions. At several points, a cylindrical tube with fixed radius was imagined, but this was only a handy visualization of the condition where all velocities are zero in planes normal to x. The *use* of the equations in Chapters 13 through 16 is to describe conditions close to a planar interface of large extent. If the ratio (distance from interface)/(breadth of a planar portion of interface) is small, behavior is as if the interface were infinitely extensive, and it is to this condition that the equations apply.

Deformation of a cylinder with no diffusion

As a step toward understanding behavior around an inclusion, we now consider a long cylinder of one material embedded in an unlimited extent of a second material. The axis of the cylinder is taken as the y-direction and we continue to assume that everything is uniform in this direction: all properties and all behaviors are uniform along y and all velocities along y are zero. But in xz planes we now see a circular cross-section as in Figure 18.1 instead of just a planar boundary.

As regards stress state, let this be uniform throughout the host material except insofar as the inclusion causes variation; let the remote stress state have principal compressive stresses σ_{xx} and σ_{zz} with σ_{zz} larger. To start, we make the same assumptions as in Chapter 13, namely that the two materials are uniform and of the same chemical composition, differing only in viscosity; and let the material of the inclusion be stiffer.

Because of σ_{zz} being larger, the entire assembly will at any moment be shortening along z and elongating along x, and if the cross-section is circular at the moment we inspect it, it will be elliptical at all later times. An impression of how deformation proceeds is given in Figure 18.2, which shows how a grid would look at a later time if it had been a square grid at the moment when the inclusion was circular. Points to note are the deformation inside the inclusion (uniform and slight), the deformation at all points far from the inclusion (uniform and greater), and the fact that deformation in the host just outside the inclusion is less than at remote points; each of the three

187

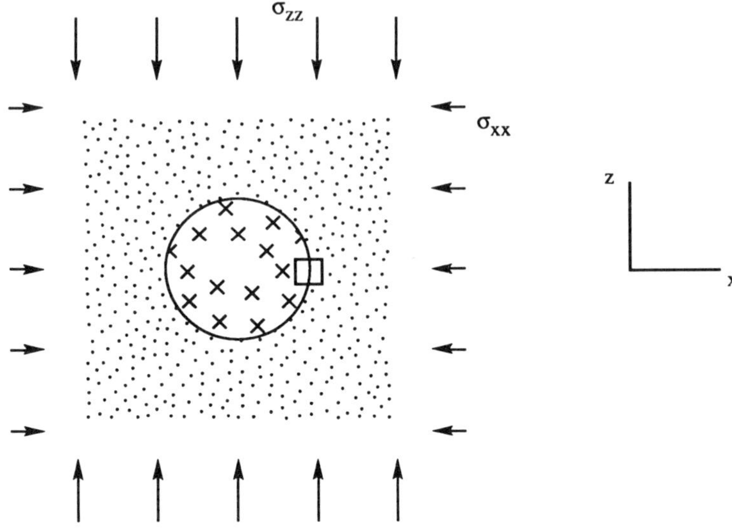

Figure 18.1 Cross-section through a cylinder of stiff material embedded in an infinite extent of less stiff material. The stress state in the host is homogeneous but anisotropic at all points, except insofar as it is modified by the inclusion.

stippled cells is closer to being square than host-cells farther from the inclusion.

The accompanying stresses are shown in Figure 18.3 as they would be if the material had no capacity whatever for self-diffusion. Inside the inclusion, σ_{zz} is uniform and high, and σ_{xx} is uniform and low; far from the inclusion, again stresses are uniform but the difference $\sigma_{zz} - \sigma_{xx}$ is smaller; close to the inclusion at its east point, σ_{zz} and σ_{xx} are both low and with an even smaller

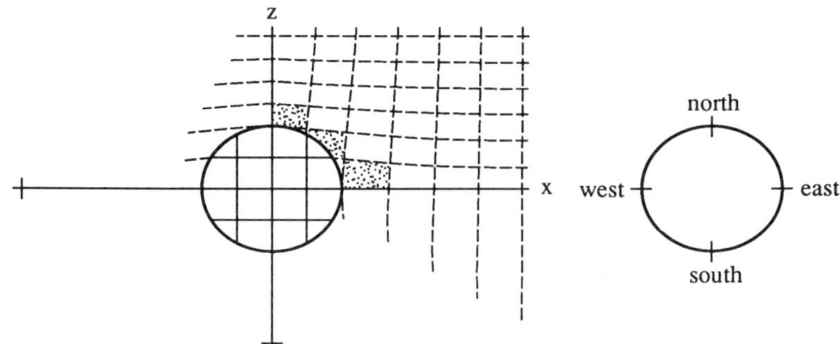

Figure 18.2 (a) A state of strain resulting from the stress state in Figure 18.1 in absence of self-diffusion. The strain rate inside the inclusion is uniform, and it is uniform again at points remote from the inclusion, with different principal values. (b) An arbitrary but convenient system for designating points around the periphery.

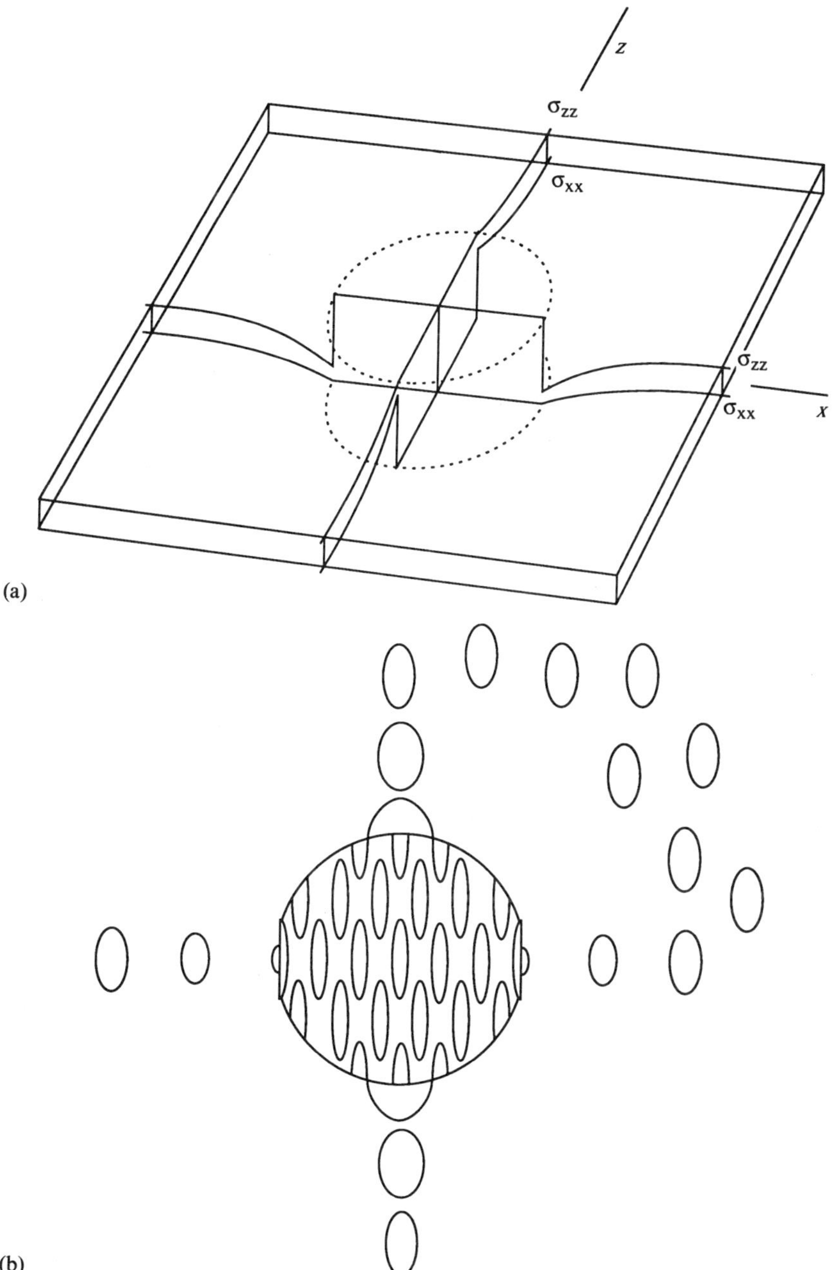

Figure 18.3 Stress magnitudes in and around the cylindrical inclusion, in absence of diffusion. In (a) stress magnitudes are shown by means of profiles drawn on lines through the inclusion's centerline along the x-direction and the z-direction. In (b) stress states are shown by means of ellipses. As with the strain state, conditions are homogeneous inside the inclusion, and again homogeneous with different principal values at points remote from the inclusion.

difference than at remote points; close to the inclusion at its north point, σ_{zz} and σ_{xx} have the same small difference but here both magnitudes are high. To compare the east point with Figure 13.1b, one has to imagine distances up the stress axis to stay fixed but distances across the page to be enormously enlarged; then the curvature of the interface would be unnoticeable and the gradients as σ_{zz} and σ_{xx} increase away from the interface would be flattened. Close to the east point, the stress state in Figure 18.3 would be seen as like that in Figure 13.1b.

It has been emphasized, particularly in Chapter 11, that if a material can creep it can self-diffuse; although for some purposes Figure 13.1b is an adequate approximation, there must be a scale on which profiles more like Figure 13.1c are seen. Thus we shall wish in due course to allow for self-diffusion and for the mechanical and chemical effects that are entailed. But even without self-diffusion, already a complication appears because of the fact that inclusion and host are likely to have different densities and thus different molar volumes.

Change of phase. It is to be recalled that in Chapters 13 through 16 the pattern has been to pose a question as follows: "Suppose that at an interface equilibrium exists at hydrostatic pressure P; then σ_{xx} is kept equal to P and equilibrium is disturbed by changing only σ_{zz}; what processes begin to run?" etc. A consequence of this approach is that we have not paid attention to processes brought on by making σ_{xx} different from P except by references to δp in part of Chapter 16. With a cylindrical inclusion, however, these processes must be considered, as follows.

There are different possible ways of reaching the stress state shown in Figures 18.1 and 18.3: one might establish an equilibrium state at pressure P_1 and then, keeping $\sigma_{xx}^{\text{remote}}$ equal to P_1, change $\sigma_{zz}^{\text{remote}}$ to a higher value; or one might establish equilibrium at P_2 and then, keeping $\sigma_{zz}^{\text{remote}}$ equal to P_2, change $\sigma_{xx}^{\text{remote}}$ to a lower value; or of course there are other possibilities. But all result in compressive stresses normal to the interface at the interface that are in general not equal to any pressure at which host and inclusion are in equilibrium. If host and inclusion have different molar volumes, the nonequilibrium pressure will favor one or the other, and change-of-phase processes will begin to run for this reason.

The easiest situation to imagine is as follows: let host and inclusion be in equilibrium with each other at some hydrostatic pressure P; then let the situation be disturbed by increasing $\sigma_{zz}^{\text{remote}}$ above P and lowering $\sigma_{xx}^{\text{remote}}$ below P by equal amounts. In this procedure, the portions of interface on $45°$ diagonals remain in equilibrium because the normal stress on them remains equal to P. Suppose the inclusion is denser as well as more viscous than the host: then at the east point, where the stress normal to the interface is reduced below P, the host will be favored, material of the inclusion will tend to change phase, and the interface will migrate inward. Conversely at the north point, the stress normal to the interface is greater than P, the inclusion is favored, and the interface will tend to migrate outward. In this condition, deformation and change of phase have opposing effects on the geometry of the interface; given the right properties, one can even make the inclusion elongate itself north–south by increasing the north–south compressive stress. But obviously if the inclusion is less dense than the host, the change-of-phase tendencies are reversed; they now affect the outline of the inclusion in the same way as the deformation; it becomes shorter north–south and longer east–west for both reasons.

Chemical consequences similarly can be of one kind or the opposite. In a situation where host and inclusion are both of formula (A, B)X with the inclusion richer in A, the last process mentioned—inclusion *growing* at the east point—will make the east point a sink for A and a source for B. But if the inclusion is poorer in A, the east point becomes a source for A, and so on. There are eight possibilities according as the inclusion is stiffer or less stiff, denser or less dense, and richer or poorer in A than the host; which of the eight make the east point a source for A and which a sink can be worked out in the manner already begun. Then again, if AX has a larger molar volume than BX, a sink region for A will gain increments of compressive stress by the change of phase, whereas if AX has a smaller molar volume, a sink region for A will have compressive stress reduced. The stress excursions shown in Figure 18.3 can thus be either reduced or enhanced by the chemical changes they provoke; as σ_{zz} and σ_{xx} diverge from the initial magnitude P, there can be positive or negative feedback from the change-of-phase effects.

Deformation of a cylinder with diffusion

It is convenient to refer to Figure 16.6, containing arrows to represent self-diffusion of the material and slabs to represent wholesale change of phase. The device was used in Chapter 8 of considering a short interval of time and, for the designated interval, dividing atoms into stayers and diffusers. In such an interval, diffusers can cross the interface at some comparatively high speed but the interface can also move through the stayers at a lower speed; the whole lattice or framework of stayers can change phase. It is this second process, indicated by slabs in Figure 16.6, that has been discussed in the preceding section.

When we come to add effects related to diffusion, as represented by the horizontal arrows in Figure 16.6, the reader may feel that there are more things going on at one time than one can keep track of; the number of simultaneous effects becomes bewildering. But it will be argued that each in itself remains intelligible; a long list of simple items is not *harder to understand* than a short list. The possibilities indicated in Figure 16.11 all range between one simple special case and the other, and treating a cylindrical inclusion involves no additional *concepts*. To actually calculate all the interactions is formidable, but listing the effects one would like to include is not so intimidating.

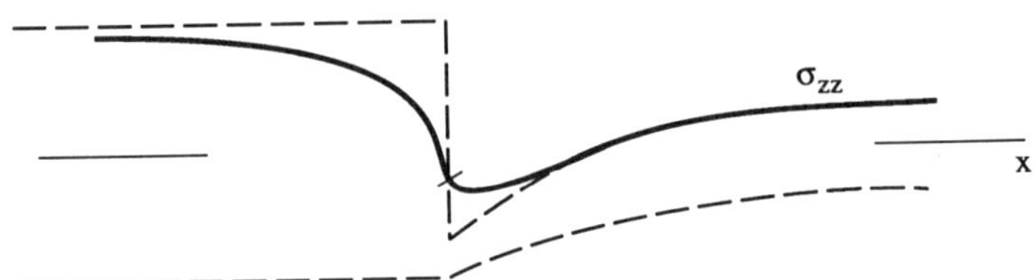

Figure 18.4 The effect of diffusion on the profile of compressive stress, assuming perfect coherence at the interface. (The profile is based on unpublished equations by Sharon Finley.)

Introducing the idea of self-diffusion had the effect in Chapter 13 of replacing Figure 13.1b by 13.1c. The east point in Figure 18.3 resembles Figure 13.1b as just noted, and a modification for self-diffusion yields Figure 18.4, resembling Figure 13.1c. (If the horizontal scale of Figure 18.4 were greatly expanded, the portion close to the interface would look very much like Figure 13.1c.)

The profiles discussed so far are those that would be expected if there were perfect coherence between the inclusion and its host. Alternatively, suppose that inclusion and host are not coherent but are separated by a thin film of weak material: then Figure 13.5 is the pattern to follow rather than Figure 13.1c. A feature of this figure is that, material 3 being of negligible viscosity, σ_{zz} and σ_{xx} inside the stiff material become equal to each other at the stiff material's boundary; in the inclusion, they are equal and high at the north point, equal and low at the east point, and equal but with intermediate magnitudes at other points around the rim. The two surfaces representing σ_{zz} and σ_{xx} inside the inclusion join to form an interesting closed shape as in Figure 18.5.

(A side issue here is as follows: it was just supposed that a weak material exists at the east point at a low compression σ_{xx}, and that a weak material exists at the north point at a high compression σ_{zz}. The continuous line shown in Figure 18.5 between these points suggests a continuous film of weak material, and if such a film is continuous one may ask about flow of the weak material from the north point to the east point. The issue is discussed under the heading "Wet granular aggregates" later in this chapter. To avoid its becoming a distraction at this point, let us imagine a series of little barriers so that the weak material can exist at different pressures at different points around the inclusion's boundary. It is agreeable that there is a precedent for this in the work of Willard Gibbs.)

Returning to the main theme and comparing Figures 18.3 and 18.4, we see that just inside the east point of the inclusion, σ_{zz} is diminished by self-diffusion or, in other words, σ_{zz} does not need to climb so high where outward self-diffusion makes a contribution to shortening along z. Similarly, just inside the north point, σ_{xx} does not need to drop to such a low magnitude where inward self-diffusion makes a contribution to elongation along x.

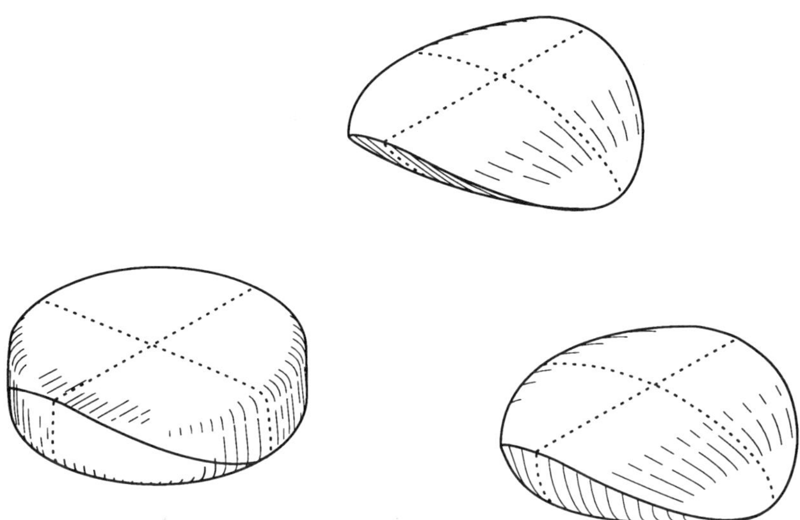

Figure 18.5 Compressive stress in the inclusion, assuming no coherence at the interface. To give a condition of no coherence, a pocket of fluid is imagined at every interface point; in the fluid $\sigma_{zz} = \sigma_{xx}$ as in Figure 13.5. Then surfaces representing σ_{zz} and σ_{xx} in the inclusion meet in a line at the interface. The shape of each surface is fixed by the ratio of the material's characteristic length to the radius of the inclusion.

Considering the overall flux of material through the region affected by the inclusion, this is greater when diffusion is taken into account. On the other hand, the rate at which the inclusion changes ellipticity is slightly less. In terms of velocity fields, the strain-rate field by itself is smaller when diffusion operates than when it does not, but the sum of strain-rate field plus the velocity field for the self-diffusive movements is greater than the strain-rate field in the no-diffusion case.

Materials of formula (A, B)X

As in earlier chapters, the question taken up is as follows. Suppose an equilibrium state exists under hydrostatic pressure P and then the stress state is made nonhydrostatic: in what ways is equilibrium upset, what processes begin to run, and can any steady state be recognized?

From the previous section we suppose that (1) the stress state in the inclusion becomes not only nonhydrostatic but also nonhomogeneous; (2) inclusion and host begin to change shape; (3) material begins to change phase at the boundary by self-diffusion, adding to the inclusion at sites of high compression (the north and south points in Figure 18.2) and escaping from the inclusion at sites of low compression. These changes are additional to any change of phase brought on by the interface normal stress departing from P, as discussed in the no-diffusion section. The material also, of course, has ability to change composition, by interchange of atoms of A and B. Let us suppose that, at points remote from the inclusion, the stress difference $\sigma_{zz} - \sigma_{xx}$ is kept steady through time, so that the strain rate is steady as well. Then possibilities are first that the A:B ratio becomes nonhomogeneous, and second that steady fluxes of A one way and B the other way develop. The change of phase at the interface is comparable with that in Chapter 16; as long as change of phase continues steadily, the interface is a steady source of either A or B and a sink for the other. Consequently, close to the interface, a gradient in concentration of A, for example, and a flux of A can be maintained with very little change through time, as long as the inclusion is large in extent in comparison with the volume of material affected by these interface processes.

In Chapter 15, it was proposed that, for a material of formula (A, B)X, the most useful point of view is to regard this as comprising a matrix or substrate whose composition cannot change, through which a constituent or additive of small volume-fraction can migrate that accounts for the change in composition. It was further proposed that, as a first approximation, the second should be considered sensitive only to the field of mean-stress magnitudes, whereas the first should be considered sensitive to the difference $\sigma_{max} - \sigma_{min}$ at a point as well as to the mean-stress magnitude.

If we use the approach just described, we expect the stress field in the substrate to be of the type shown in Figures 18.4 and 18.5; then the mean stress forms a single surface as shown in Figure 18.6. What effect will such a nonuniform pattern of mean stress have on the distribution of components A and B?

Obviously both components will tend to migrate from high compression to low compression, but because of the stoichiometry constraint (i.e., the powerful restraining effect of any slight departure from electrical neutrality)

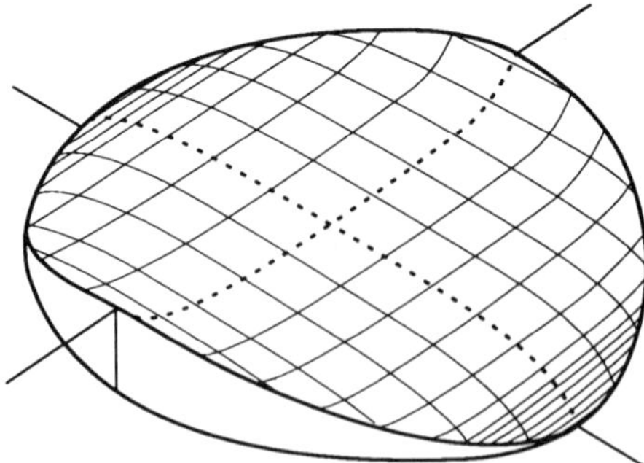

Figure 18.6 Mean stress magnitudes in the inclusion, as represented by a single mean-stress surface. Except for the effects of diffusion, the conditions specified give plane strain; hence the mean stress is close to $(\sigma_{xx} + \sigma_{zz})/2$.

they cannot both do so. Suppose, as before, that the endmember AX has a larger molar volume than BX: then component A will move down the mean-stress gradient and component B will move against it until nonuniformity of composition counteracts the volume difference. (Or in terms of the fictional additive *a*, the additive will move until it is sufficiently concentrated at the low-stress end for its concentration gradient to just counteract the mean-stress gradient.)

The upshot is that we expect a lower concentration of A just inside the north point, an almost uniform concentration of A through much of the inclusion's interior and a higher concentration of A just inside the east point. We have in mind the idea that fluxes are running; hence we do not suppose that the stress effect and concentration effect exactly balance all across the inclusion. But, by continuing the consideration of orders of magnitude from Chapter 16, it can be shown that deviations from a static balance are only small: hence an important effect of making the uniform stress state remote from the inclusion nonhydrostatic is to generate a nonuniformity of composition inside the inclusion that is close to the particular nonuniformity that would exactly compensate for the nonuniform mean stress. If C^A and C^B are both of the order of magnitude 0.50, deviations due to the stress state might be 0.01 (e.g., $C^A = 0.49$ at the north point and 0.51 at the east point), whereas deviations associated with the fluxes might be only 1/100 as big, with order of magnitude 0.0001.

Fluxes. The fluxes driven by the nonhydrostatic stress obviously depend on both the chemical and mechanical characteristics of the system. For the sake of a specific example, let us continue to suppose that the inclusion is stiffer than its surroundings, σ_{zz} is greater than σ_{xx}, AX occupies more space than BX, and that in equilibrium the inclusion has a higher A:B ratio than the host. Then conditions at the east point are as in Figure 16.9a: the interface is retreating into the stiffer material and liberating an excess of A. Conversely, at the north point the interface is advancing into the host and A is being absorbed by the change of phase. Let us suppose that the density contrast between host and inclusion is such that change of phase from this cause simply adds on to the diffusion effects. What then can be inferred about the concentration pattern of component A?

In the previous section it was noted that, for components A and B to stand still in presence of the nonuniform mean stress field, the concentration of A needs to be higher at the east point; now we see that the concentration at the east point needs to overshoot this condition a little if it is to drive the fluxes that accompany the change of phase. But, of course, if the chemical relation were of the opposite kind, with the host having higher A:B ratio in equilibrium, the retreating interface at the east point would absorb A, the concentration of A there would be a little less than the concentration for static balance, and there would be a small flux of A *toward* the east point.

Summary

If an inclusion has a higher viscosity but is coherent with its host and the assembly is compressed, then if the material is taken as chemically inert, the stress field takes the form shown in Figure 18.3. It is more realistic to allow for diffusion and to envisage stress profiles as in Figure 18.4.

If, instead of being coherent, the interface is a layer of non-communicating pockets of weak material, the stress field takes the form shown in Figure 18.5.

In the condition of Figure 18.5, material is likely to change phase at the interface for two reasons. First, the hydrostatic pressure at an interface point will in general not be the pressure at which host and inclusion are in equilibrium, and second, the stress gradient leading to the interface point from the inclusion's interior will drive material through the phase-change by diffusion. The two effects are independent and may supplement or oppose each other.

If the two phases both have formula (A, B)X, the deformation and all changes of phase involve the substrate of atoms of X. Superimposed on the movement of the substrate, there is stoichiometric redistribution of A and B, and this has a static aspect and a flux aspect.

Components A and B will tend to take up a static distribution compatible with the field of mean stress, Figure 18.6. The component of lower molar volume will tend to have higher concentration at sites of higher mean stress.

A static balance of concentration versus stress will not be achieved exactly. Departures from exact balance will exist that result in fluxes of A and B that bring A and B to or from the interface at rates compatible with the rate of change of phase there.

Orders of magnitude from Chapter 16 suggest that, for materials of the type discussed there, variation in A:B ratio will be dominated by the static effect, with only small deviations related to the concurrent fluxes. Variation in stress will be basically as in Figures 18.4 or 18.5, but modified on account of interchange of A and B in the same manner that Figure 16.11c modifies Figure 16.11a.

Wet granular aggregates

Statements in preceding sections about compounds of type (A, B)X have all been made strictly on the assumption of stoichiometry; that is to say, we have assumed that a necessary condition for component A to move is that component B should move in the opposite direction on an atom-for-atom basis. But the stress states in Figures 18.5 and 18.6 are controlled mainly by the properties of the substrate and are not much affected even if the presence of a third component alters the stoichiometry constraint. A third component of considerable interest is hydrogen: if X is an oxy-anion instead of a sulfide, then wherever there is water there is the possibility of hydrogen acting as a cation, with interesting results. The following remarks are set in a geological context but obvious parallels in man-made materials will occur to the reader.

Before considering the presence of a small quantity of water, we consider a polycrystalline aggregate of an oxide or oxy-salt. The stress state immediately becomes infinitely complicated in detail, but if on average σ_{zz} is greater than σ_{xx}, it continues to be true that in general at north and south points of grains the mean stress will be high and at east and west points it will be lower (compass directions are used still in the manner of Figure 18.2). Instead of a featureless extensive host, a grain is surrounded by an aggregate of grains like itself; and instead of the interface being a uniform film of "material 3," the interface is rather irregular and variable. At some points we may have an almost coherent junction of grain with neighboring grain, at some points an approximation to a pocket of liquid water and at some points perhaps a less easily described material of transitional character. The details of the material do not greatly affect the following suggestions.

The situation in view resembles an (A, B)X inclusion in an (A, B)X host in many ways, but has this additional feature: it is now possible for both A and B to migrate through the substrate X and down the mean-stress gradient, as long as hydrogen ions migrate up the gradient to maintain charge balance. Being of such tiny volume, hydrogen ions do this quite readily. Upon arrival at a high-stress site that has been vacated by species A and B, such as the north point of a grain, a hydrogen ion is likely to combine with some of the oxygen there to form water. Examples of possible reactions are

$$\text{Mg}_2\text{SiO}_4 + \begin{pmatrix} 2\text{H}^+ \text{ in and} \\ \text{one Mg}^{2+} \text{ out} \end{pmatrix} = \text{MgSiO}_3 + \text{H}_2\text{O}$$

$$\text{CaCO}_3 + \begin{pmatrix} 2\text{H}^+ \text{ in and} \\ \text{one Ca}^{2+} \text{ out} \end{pmatrix} = \text{CO}_2 + \text{H}_2\text{O}$$

Any water so formed also tends to migrate from higher pressure to lower pressure, and it is here that we come to a resolution of an earlier difficulty, as follows.

The simplest mechanical situation for a cylindrical inclusion involves a thin layer of fluid at the interface that is at hydrostatic pressure at every point. But if such a layer were continuous, the fluid would quickly flow away from the high-stress point; earlier in this chapter, a rather unrealistic set of imaginary barriers was introduced to get over this difficulty. With a real granular aggregate of an oxy-salt, no such barriers exist and fluid does flow away from high-stress points; but the supply is continually replenished by hydrogen ions migrating up the gradient to those points and decomposing the oxy-salt upon arrival. Another way of describing the processes is to say that hydrogen goes around a loop, traveling up the pressure gradient as ions (protons) and down again as water molecules; oxygen atoms travel down-gradient as water molecules, and species Mg, Ca, Fe, etc., travel down as cations. The overall effect is to transport MgO, etc., from north points to east points; it is a mechanism by which the aggregate as a whole can change shape; it is also a mechanism that ensures that, however long the stress field is maintained, there is perpetually a certain amount of mobile fluid present at the high-stress site and in process of escaping. It is the action of hydrogen that maintains this paradoxical situation.

The type of nonuniform composition discussed in this chapter is recorded by Ozawa (1989) in naturally occurring crystals of spinel. The stress state at the time when the anomalous compositions developed is not known, but the crystals described are good examples of the type of heterogeneity predicted in this chapter.

19

REVIEW OF STRATEGIES

This chapter has two purposes, one limited and one more extensive. The limited purpose is to reconsider simply binary materials like ideal alloys, where two components A and B or copper and tin mix with each other in any proportion without being encased in a matrix of a third species X.

Considering such materials prompts a review of basic strategies for nonhydrostatic work, and the second purpose is to complete that review. A fundamental question is: granting that direction-dependent potentials can be defined and used, does any advantage derive from the practice? We find that over a wide range of situations the practice is merely an option, and yields no results that cannot be gained by other means. But there is a class of situations, probably with analogs in the real world, that seems to require that practice. At least, these situations make one hesitate to deny axiomatically that a chemical potential can be direction-dependent.

Simple binary materials

The materials in view contain just two atomic species A and B, as in an ideal alloy with no microstructure. Influences on the movement of the two components are shown in Figure 19.1. The upper part of the figure shows six boxes; each box represents the response of one of the components to one of the local gradients; the three gradients distinguished are of composition, mean stress, and stress deviator, i.e., for a plane $\mathbf{n}$, the difference (σ_{nn} − mean stress). The rest of the figure, and indeed the whole of the book, is concerned with how to amalgamate these responses.

One way of combining the responses is shown across the top of the figure. We form a sum of responses or force–flux relations for component A by itself, do the same for component B by itself, and then face the question: how will A and B interfere and make their combined response different from the simple sum of their in-isolation responses? This method can employ eqn. (12.8):

$$\text{Strain rate of A} = K^A(\partial^2/\ldots)\mu^A$$

The method was used in Chapter 14 and quickly showed the general form of the result there sought, but left open the matter of properly quantifying the interactions of A and B.

A second way of combining the six responses is shown in the middle of the figure. The method uses the idea from Figure 15.3 that two separate

199

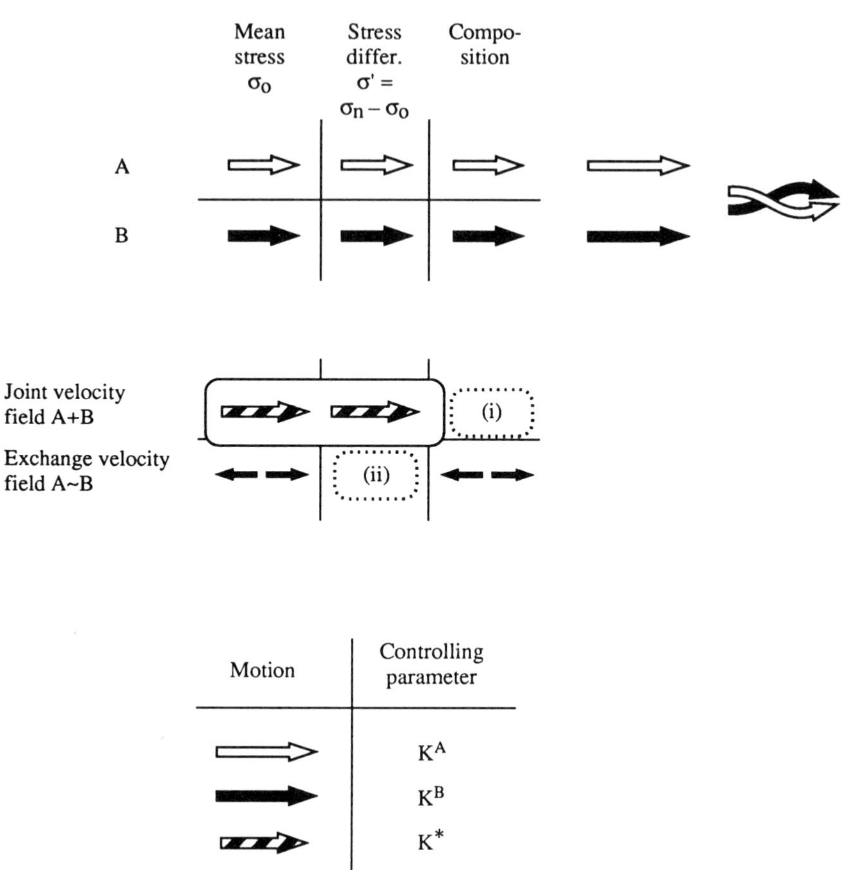

Figure 19.1 A diagram of possible strategies. To combine the six effects represented by arrows in the top left, one may proceed across the page or down the page. See the text for a more complete discussion.

velocity fields for species A and B can be described as the sum of a joint field and an exchange field. Compounds of formula (A, B)X were used to illustrate this approach; they have the advantage that the joint velocity field is concrete and easy to visualize, being just the velocity field of the species X. The field of exchange velocities can be chosen so that, by itself, it totally accounts for whatever changes of composition occur; then the joint field describes the movement of a material of uniform composition so that box (i) in the diagram becomes vacant.

The reason that box (ii) also becomes vacant lies in the exchange process; if we treat atoms as spherical and the substrate is isotropic, all effects of such an exchange must be isotropic, so that effects linked to the nonhydrostatic part of the stress must be zero. It is the emptiness of these two boxes that gives the second approach a definiteness that is lacking at the top of the page.

Consequences of the second approach are that we usually do not need eqn. (12.8) in its full form. For the joint field, eqn. (12.7) is general enough,

$$(\text{Strain rate})_{nn} = K^*(\partial^2/\ldots)\sigma_{nn}$$

—we do not need to generalize from stress σ to potential μ. And for the exchange field we need only the isotropic terms:

$$\text{Isotropic strain rate } e_0 = K^a(\partial^2/\ldots)\mu_0$$

In a sense we treat the material as a gas dissolved in a continuum. The continuum part is chemically inert but responds in the classical way to the total stress field, both the mean stress and the deviatoric stress and their gradients; it supports the whole of the nonhydrostatic stress components. The gas part has a completely different mobility coefficient, and the idea of its being affected by the deviation of the stress state from hydrostatic is rejected.

One more way of expressing the advantage of the pair K^*,K^a over the pair K^A,K^B is thus: with K^* we clearly need products of both kinds $K^*\,\partial^2\sigma_n/\partial x^2$ and $K^*\,\partial^2\sigma_n/\partial\theta^2$, while for K^a we clearly need only $K^a\,\partial^2\sigma_0/\partial x^2$. If we try to work with K^A and K^B, what to do about the nonhydrostatic stress components is far less clear.

The conclusion is that for simple binary materials, to work with K^* and K^a, with a joint velocity field and an exchange velocity field, is the preferred approach, as it is for materials of type (A, B)X. It seems that even when no stoichiometry constraint operates, to use an atom-for-atom scheme for describing changes of composition is an efficient way to work.

The question naturally follows: How might one try to determine K^* and K^a experimentally? Here Chapter 16 provides a start; referring to Figure 16.8d, to depart from equilibrium along path (1) would permit K^a to be measured, while to depart from equilibrium along path (3) would yield K^*. Figure 16.11 is a reminder that these two paths give special or endmember behavior, dominated by one coefficient or the other without the complications of a mixed effect.

Questions of scale

The preceding section probably overstates the advantage of the second approach (Figure 19.1 center) over the first approach (Figure 19.1 top). In some sense the second approach involves treating the material as comprising two endmembers, of which one is unable to show any continuum attributes and the other unable to show any compositional variation. Treating simple endmembers is, of course, a widely used and effective means of building up understanding, but the practice has two forms. In one, the study of endmembers leads simply and naturally into study of intermediate states; in the other, the assumptions made in order to pinpoint the endmembers hinder study of intermediate states. Enquirers may find that they have locked themselves into discussion of only endmembers, and developed patterns of thought that prevent them from making progress with intermediate states. The second approach described seems to have this undesirable side effect.

These general remarks can be made more concrete by considering the matter of scale; what is true when the world is viewed on only one scale may not be true when two scales of phenomena are in view at one time. We are concerned in particular with the emptiness of boxes (i) and (ii) in Figure 19.1. To say the boxes are empty means that one cannot prompt an (A, B)X sample to change shape by driving the substitution of B for A, nor can one prompt a change of composition by a homogeneous deformation. For isotropic materials, these statements are true as long as the composition field and the stress field are viewed on the same scale, but it is easy to find counterexamples if two scales are used.

Two illustrations are given in Figure 19.2. Each half of the figure shows a representative part of an extended sample; the squares shown can be considered cross-sections through long square prisms normal to the page. Two components are represented by small circles and larger circles. In diagram (a), some prisms are rich in component A and some in component B: the chemical heterogeneity drives fluxes through the intervening neutral

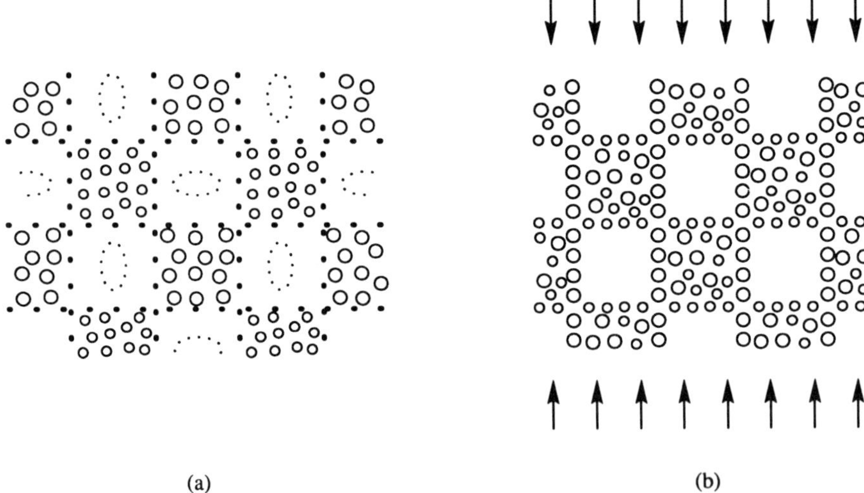

(a)   (b)

Figure 19.2 Cross-sections through idealized inhomogeneous assemblies. Each assembly is built from long square-ended prisms that are joined to form a continuum. Large circles and small circles mark regions rich in one component or the other; unpatterned regions are of neutral composition. In (a) the stress state is hydrostatic in gross but nonhydrostatic locally on account of chemistry-driven diffusion. In (b) the chemistry is uniform in gross but nonuniform locally on account of stress-driven diffusion.

prisms that will tend to cause these to change shape, in the manner shown by the ellipses or in an opposite manner, according to the transport characteristics of the two traveling species. Viewed on a large enough scale, the material is chemically homogeneous and of stable shape, but if, as shown, it is chemically heterogeneous on the scale of the prisms, it can contain even smaller subsamples that are in course of changing shape more or less homogeneously for chemical reasons.

The other part of Figure 19.2 is the inverse situation, where the ornamented squares are all of similar composition, and of uniform composition until deformation is imposed. Then if the unmarked squares are a material that permits rapid diffusion, sorting of components A and B will occur in the manner indicated by the rows of one component or the other; this is a highly stylized version of effects like those discussed in Chapter 18.

The diagrams discussed are highly artificial and would not justify any conclusion about the behavior of real materials if it were not for the fact that a material has a characteristic length, as discussed in the next section.

A material's characteristic length and microstructure

As always, we need to separate the real from the imaginary. A material's viscosity N is real, its coefficient for self-diffusion K is real, its characteristic length R in nanometers or micrometers is a real attribute of the material, with $R = 2(NK)^{1/2}$ as in Chapter 12. By contrast, a quadrant arc of radius R is imaginary and even more so is a swarm of such arcs, as in Figure 11.5. Then what microstructure is real, when R equals say 200 nm?

In a polymer we might get one answer and in a ceramic material we might get another. We emphasize in this section what is *not impossible*. It is not impossible for a material to have a microstructure of the type illustrated in Figure 19.3. Again the figure is highly stylized but it is not impossible for a real material to comprise less diffusive islands among which are distributed

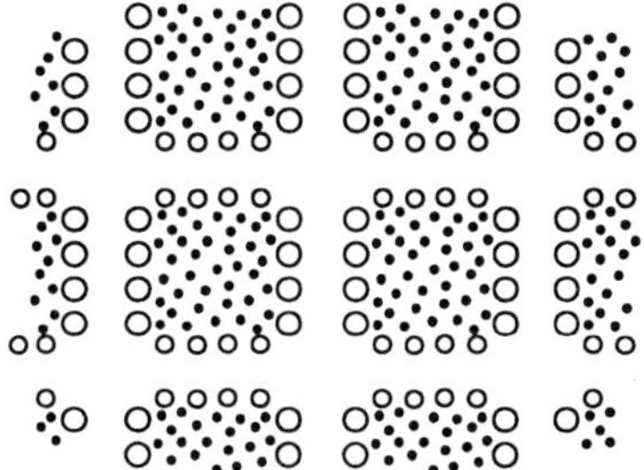

Figure 19.3 An idealization of a material where interdiffusion of components can occur with limited facility at all points but with greater facility along identifiable high-diffusion channels. In such a material, nonhydrostatic stress can produce a difference in composition at the channel wall between channels of one direction and channels of another direction.

more highly diffusive channels. It is not impossible that a material's characteristic length is in some instances linked to the geometry of such channels. It is not impossible that a material responds to nonhydrostatic stress by material diffusing—to some extent pervasively so that every point in the material is affected, but to some extent selectively, with the channels as favored conduits. It is not impossible for the characteristic length R to have some basis in a real network of channels of this type, and for deformation to occur primarily by movement of material through space along them. Then it is not impossible, on average over the sample, for the chemical environment at one end of a nonhydrostatically-driven diffusive journey to be different from the chemical environment at the other end; east–west channel walls can be chemically different from north–south channel walls.

The preceding six sentences are about real materials. We now turn to the question: if we wish to imagine an ideal material whose behavior follows equations and yet resembles real materials in selected ways, what properties should we assign to the imaginary material? First, we should imagine it to be free of microstructure so that it obeys the same equations at every point, but, second, we should assume that $\partial^2 \mu^A / \partial \theta^2$ might *not* be simply the product $V^A \, \partial^2 \sigma / \partial \theta^2$. It is possible for a real material to have direction-dependent chemical attributes, and these require the potential μ to be a more complicated function of orientation θ in any idealization that is to represent the real material.

Regarding what is possible and what is not possible, we can come at the distinction from another point of view. Figure 19.4 shows an extensive sample of a material with two components A and B, and two lines through the sample along which profiles of the concentration of A can be measured. If both profiles represent the material without bias, they must have the same mean value; it is not possible for the mean values of two unbiased profiles

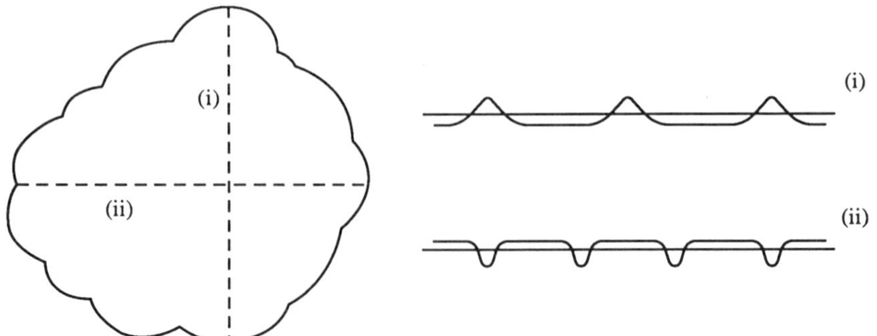

Figure 19.4 Profiles of composition along two lines through a sample as in Figure 19.3. If the sample is homogeneous in gross, the mean values along the two profiles must be equal, but small-scale attributes such as peak heights need not match.

to be different. But it is possible for the peak heights, etc., to be different, and it is also possible for a material's behavior—for example its reactivity—to sample the profiles selectively. A material's free surface, for example, might tend to coincide with a peak in profile (i). Applying a spatially uniform nonhydrostatic stress cannot change the bulk composition of the sample nor change the mean height of profile (i) or (ii), but nonhydrostatic stress might change the peak heights. The material's chemical behavior then becomes direction-dependent and open to change by changing the nonhydrostatic part of the stress state σ_n' or $\sigma_n - \sigma_0$.

Summary

A binary material's behavior can be described by giving a velocity field for the joint behavior of both components plus a velocity field for the exchange of one for the other. In many circumstances this effectively separates questions of mechanics from questions of chemistry; in particular, nonhydrostatic stress affects only the joint field, while gradients of composition through space affect only the exchange process. In such circumstances, boxes (i) and (ii) in Figure 19.1 are empty and eqn. (12.8) relating the strain rate e^A to the potential μ^A is exactly equivalent to eqn. (12.7) in terms of the stress σ.

On the other hand, there are real circumstances in which nonhydrostatic stress affects a material's chemical behavior. In these circumstances, eqn. (12.8) is needed, eqn. (12.7) not being sufficient. Boxes (i) and (ii) in Figure 19.1 contain nonzero effects and behavior is *as if* the concentration of a component at a point in the material could vary with direction. To construct theory using an axiom that concentration cannot vary with direction is unduly restrictive; such an axiom prevents the theory from effectively describing some effects that are possible in real materials.

A separate complication is that the substrate may be anisotropic; this possibility is discussed in Chapter 20.

Invariants

The following remarks use ideas that are introduced in Appendix 8C. We emphasize effects that occur or exist at a point that have no direction associated with them, as opposed to flows and gradients with respect to position, which are vectors. The purpose is to repeat the summary just given in another, more categorical, form.

Appendix 8C describes a procedure for specifying the stress state at a point by means of three invariants rather than three principal stresses; the invariants are totally free of any link to directions. The point is also made that the Laplacian operator has the same attribute; any first derivative with respect to position, $\partial/\partial x$, is a vector, and a single second derivative $\partial^2/\partial x^2$ is also linked to the direction x, but the Laplacian has no links to direction

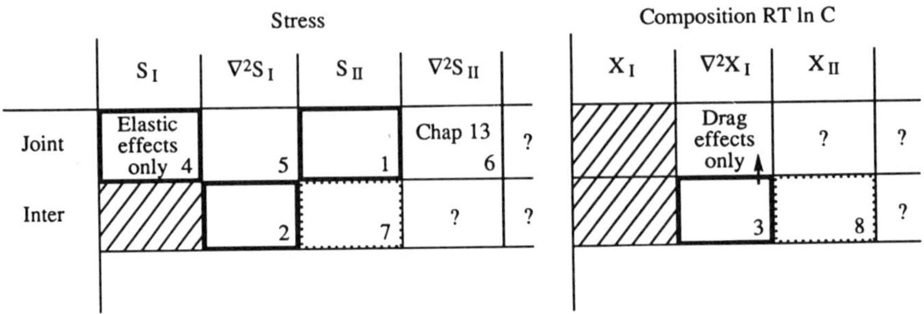

Figure 19.5 Possible effects of stress and composition listed using invariants (top row). For a two-component material, effects on the joint behavior of the two components are shown in row 2, and effects on the interdiffusion of components are shown in row 3.

whatever. If one side of an equation involves only this type of direction-free quantity, the other side must do the same. Using this idea we can redraw Figure 19.1 in a more definite form, as in Figure 19.5.

We have in view a material with two components that are moving around. As before, we separate the movements into a velocity field by which A and B exchange places and a velocity field for a chemically uniform substrate. Then at any point there are direction-free effects of exchange, such as change of composition, and direction-free effects in the substrate, such as diffusive gain or loss. The purpose of the figure is to show what influences might drive these effects; because we are focusing on direction-free effects, the influences can only be invariants or their Laplacian functions.

In Figure 19.5, the top row shows influences on the material motions:

S_I　　the first stress invariant or mean stress

S_{II}　　the second stress invariant, which measures the extent to which the principal stresses differ from the mean stress

X_I　　the first invariant of composition, which is just a component's concentration

X_{II}　　the second invariant of composition, usually taken as zero; but it is possible that nonzero magnitudes arise in materials with microstructure, as discussed in the previous section.

Laplacian functions of these are also possible direction-free influences.

Direction-free consequences or effects are, first, a rate of change of composition $\partial X_I/\partial t$ and, second, strain effects: the mean strain rate at a point (1/3 × rate of change of volume), and the rate of change of shape. These are represented by E_I and E_{II}, the first and second strain-rate invariants.

Familiar effects represented in Figure 19.5 are:

At box (1), S_{II} affects E_{II}; nonhydrostatic stress makes a material change shape.

At box (2), $\nabla^2 S_{\mathrm{I}}$ affects $\partial X_{\mathrm{I}}/\partial t$; a local high value of mean stress drives interdiffusion of high-volume components and low-volume components.

At box (3), $\nabla^2 X_{\mathrm{I}}$ affects $\partial X_{\mathrm{I}}/\partial t$; a local high value of some component's concentration also drives interdiffusion. (These three are the simplest, most well-known connections.)

At box (4), we look for a link between S_{I} and E_{I}—a mean stress producing a rate of reduction in volume. As discussed on page 50, the more familiar effect is that a mean stress produces a reversible elastic reduction in volume that does not change with time.

At box (5), $\nabla^2 S_{\mathrm{I}}$ affects E_{I}; a local high value of mean stress drives material away from the site by self-diffusion at constant composition, as well as driving the change of composition in box (2).

At box (6), $\nabla^2 S_{\mathrm{II}}$ affects E_{II}; if the stress state is more strongly nonhydrostatic at some points than at others, volume-loss by self-diffusion will not be isotropic and change of shape will occur. [Boxes (5) and (6) are the subject of Chapter 13.]

Then, coming to more speculative possibilities, at box (7), there may be an effect as in Figure 19.3: making stress nonhydrostatic might drive one component away from selected east–west surfaces and another component away from selected north–south surfaces—a chemical equivalent of piezo-electricity. A material with nonzero X_{II} would be produced.

Last, if such a material were created and then the stress was relaxed, the effective anisotropy of the material's chemistry would drive some kind of interdiffusion as the material reverted to normal; this would be an effect in box (8).

The review of boxes just made does not contain any new ideas; all the ideas mentioned are already present in earlier chapters. The purpose has been more to use the diagram as a check. Restating the ideas in terms of invariants helps to keep them separate and yet somewhat orderly, and so helps us to decide which ones we really want to believe.

20
FURTHER EXTENSIONS

The purpose of this chapter is to give attention to three directions along which more ideas could be attached, building on the preceding chapters as base. Labels or titles for the three directions are:

> unsteady behavior and elastic effects;
> the factor f;
> anisotropy.

Unsteady behavior

Throughout the preceding chapters, a highly artificial practice has been followed: attention has been focused on states where processes are occurring in the steadiest possible manner. The purpose of this chapter is to consider the question: if processes are less steady, can we still describe them concisely and predict their evolution? If we can, presumably it is by adding some terms to the descriptive equations, and we consider briefly what kinds of terms might be needed. As with turning from a single cylindrical inclusion to a granular aggregate, there is an immediate change to a vast field of complexities. The purpose of the chapter is to give just a preliminary view of how one might begin to identify possibilities.

Rheological models

The purpose is to enquire how unsteady or transient effects might occur in a system that is capable of steady behavior. For this purpose, something simpler than a chemical nonhydrostatic system can be used, as shown in Figure 20.1.

The simpler of the systems illustrated, Figure 20.1a, consists of a weight that is supported by two elements P and Q. The elements are known as dashpots; each is imagined to consist of a cylinder and piston; each cylinder is full of oil both above and below the piston and each piston has a hole. In consequence, when the element is pulled it can change length as fast as oil can slip through the piston's hole, and ideally the rate of elongation is proportional to the force pulling on the element. The fact that ultimately the piston comes to the end of the cylinder is ignored; we imagine P and Q to have as much length as we need.

In Figure 20.1a, the system is such that the two elements have to elongate

208

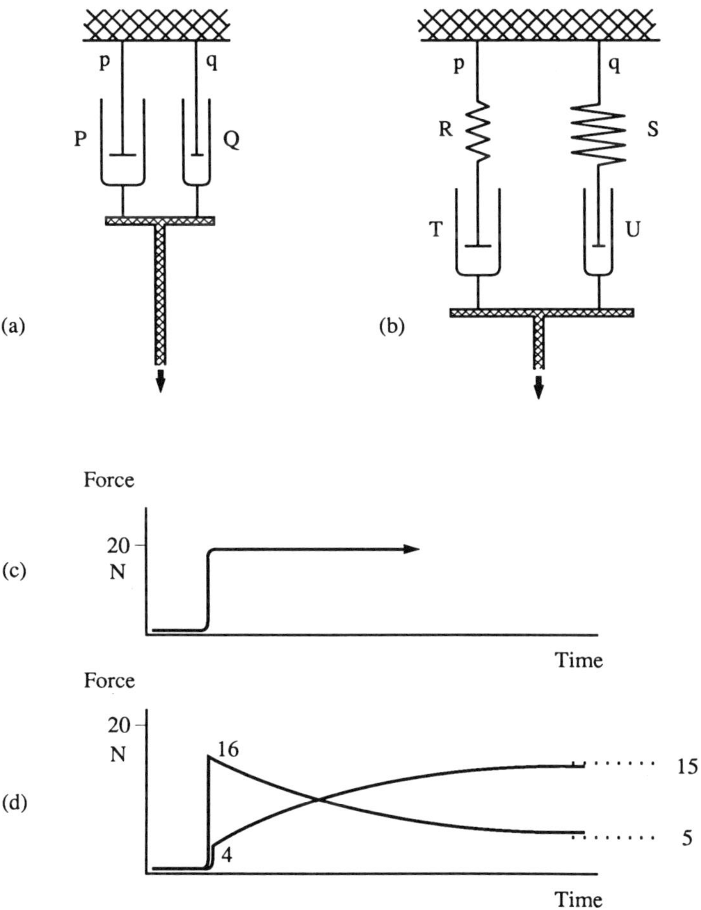

Figure 20.1 Idealized rheological systems. (a) A system with two parts; in each part, the only behavior shown is viscous (no energy recoverable). (b) A system with two parts; in each part, behavior is the sum of a viscous element and an elastic element. In both systems, the load is imposed in such a way that the displacements in the two parts are equal. (c) A possible loading history for the system in (b). (d) The loads in the two parts of the system as they would change with time. Equations and a numerical example are given in Appendix 20A.

at equal rates; but they are dissimilar and we imagine a system where, to achieve equal rates, the force pulling P needs to be three times the force pulling Q. Then if the weight exerts, say, 20 newtons we expect a force of 15 newtons in string p and a force of 5 newtons in string q.

Turning now to Figure 20.1b, we have four elements; two springs R and S have been added to two dashpots and, as the diagram is intended to suggest, spring S is stiffer than spring R while dashpot T is more resistive than dashpot U. Specifically, let T require three times as much force as U for equal elongation rates, but let S require four times as much force as R for equal elongations. We consider how stresses in the system change with time if the load of 20 newtons is applied at some instant t_0 and kept steady thereafter, as shown in Figure 20.1c.

In the first instant after loading, the dashpots have no time to respond; instantaneous sinking of the horizontal bar occurs only by elongation of the springs and, in line with the specifications above, the forces in the two strings p and q will at once become 4 and 16 newtons, respectively. On the other hand, after sufficient time (and assuming as above that the dashpot cylinders are sufficiently long) we expect to reach a state where the springs have settled down to two steady states of elongation, the continued sinking of the weight is occurring only by action of the dashpots and by now the forces in the strings p and q are 15 and 5 newtons, appropriate to the properties of T and U. The history of the forces in the two strings is as shown in Figure 20.1d; without analyzing any details, we conclude that there must be a gradual redistribution of forces, with string p taking up increasing amounts of force as the weakness of dashpot U allows the force in string q to diminish.

Before drawing conclusions from this example, let us consider the variant in Figure 20.2. All the elements of the system are as in Figure 20.1b but the nature of the overriding *constraint* is different. In Figure 20.1 the system is constrained so that the displacements and displacement rates on the two sides are always equal; by contrast, in Figure 20.2, the displacements on the two sides are not directly related but the forces in the strings p and q are

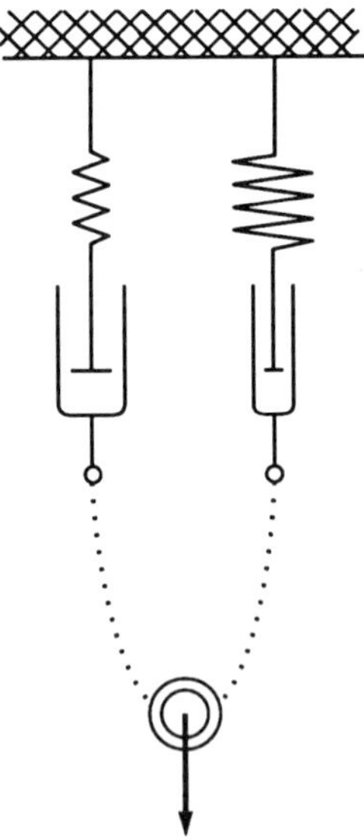

Figure 20.2 An idealized rheological system where the load is imposed in such a way that the loads in the two parts are equal but the displacements need not be equal.

always equal. (Strings p and q are linked, in fact, to constitute a single continuous string.) If we imagine the system in Figure 20.2 to be loaded with 20 newtons, of course its evolution through time will be quite different from that of the system in Figure 20.1.

The purpose of these examples is to establish the following points:

1. To specify the long-term behavior, we do not need any knowledge of the springs.
2. To specify the transient early behavior we do need knowledge of the springs (except that, as regards order of magnitude, neither force in string p or string q is likely to exceed 20 newtons).
3. To estimate the time needed for the early excess force in string q of $(16 - 5)$ or 11 newtons to diminish to half its original value, or 1/10 or $1/e$ or any desired fraction, one has to construct a differential equation and then solve it; the answer depends on the properties of both the springs and the dashpots (see Appendix 20A).
4. All of the above are standard ideas. Any system that contains elements that respond to a force F partly by a response r and partly by a response-rate dr/dt can be imitated with springs and dashpots, or alternatively with an electrical system of imagined capacitors and resistors. The technique of such simulations is well developed. In a new situation, one does not have to establish any new technique, but one does have to ask: for the present new situation, what behaviors resemble springs, what behaviors resemble dashpots, what driving agents function as forces, what kinds of constraints operate, and what is the pattern or system of links that joins these all together into a system? The purpose of Chapters 13 through 19, including the review-diagram Figure 19.1, is not to produce descriptive answers but to make us familiar with the forces, responses, and constraints in the system. Such familiarity seems to be a necessary basis. We have studied a chemical nonhydrostatic system, emphasizing its dashpots; readers who wish to consider also the effects of its springs must go on to other texts.

Experimental work

Earlier chapters contain very few references to experimental work. The reason is that relevant experiments are very difficult to perform; the distance scales are so small that keeping track of either a stress profile or a composition profile as it evolves presents great difficulties. To illustrate the difficulties, a set of experiments by A. Y. Sane and A. R. Cooper (1987) will be briefly described. These were conducted with great ingenuity, skill, and care; they provided estimates of compressive stress at sites as little as 2 μm apart and (indirectly) followed the change of these stresses through time, and yet they did not capture data that can be effectively compared with ideas from Chapter 16. (Of course, that comparison was not the objective of the

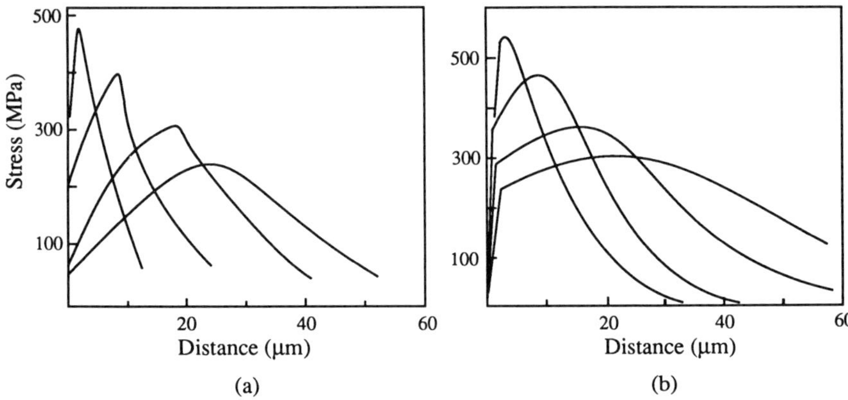

Figure 20.3 Comparison of experimental results (a) with predictions from a simple theory (b). In the theory it is assumed that change of composition affects stress but that stress does not affect change of composition. It is suggested in the main text that this is oversimple and that stress does affect change of composition, but the effect would be noticeable mainly in the first micrometer or two. Differences between (a) and (b) are certainly present, but there are several possible contributing causes so that one cannot extract definite information by inspecting the differences. (Figure 20.3a is redrawn from A. Y. Sane and A. R. Cooper "Stress buildup and relaxation during ion exchange strengthening of glass," *Journal of the American Ceramic Society*, *70* (1987) 86–89, Figure 1(D); reprinted by permission of the American Ceramic Society.)

experiments; the link is made here so as to use Sane and Cooper's experiments as an indication of what would need to be done to test Chapter 16.)

In each experiment, a sample of sodium glass with a carefully prepared surface was exposed to a potassium-rich fluid for a controlled time at a controlled temperature. Times ranged from 15 minutes to 9 days and temperatures ranged from 385 to 490°C. As potassium atoms diffused into the glass, compressive stress built up, especially in the first 10 µm of sample and especially across planes normal to the surface; see Figure 20.3a. Because the glass surface was mechanically free, no compression built up across planes parallel to the surface. No experiment directly yielded a history of stress magnitudes; but by running six similar tests and terminating them at six different times, a representative history was synthesized, and then the entire routine was repeated at a new temperature.

For analysis, let us assume that the concentration profile migrates inward as shown in Figure 16.1b with an error-function profile whose width increases in proportion to $(\text{time})^{1/2}$, and let us assume for a start that the stress state has no effect on the migration of potassium. Let us assume that the compressive stress at any point tends to rise because of this potassium concentration effect but tends to fall because of creep or relaxation of the glassy host. And let us assume that at small distances (less than 1 µm) from the surface, stress is relieved also by self-diffusion of the glass, so that right at the surface compressive stress is negligible even on planes normal to the surface. These rather coarse assumptions give Figure 20.3b and thus seem able, in principle, to account for the main features of the experimental results.

The negative conclusion is that even these rather strong stress gradients do not have effects on potassium migration that are readily detectable. For the main body of the diagrams, from 2 µm up to 30 µm, the assumption that stress has no effect on migration seems adequate to first order; stress effects, if present, are too small to be readily identified. For the extreme edge of the diagram, the stress effects must be larger—must ultimately become dominant—but here the spatial scale is inconveniently small. Thus although the experiments display very well the effect of concentration-driven migration on stress, they do not provide clear evidence about the effect of stress on migration.

Poorly correlated materials and the factor f

The factor f appears in eqn. (12.7):

$$e_{nn} = \frac{K}{3}\left[\frac{\partial^2}{\partial(R\alpha_1)^2} + \frac{\partial^2}{\partial(R\alpha_2)^2} + \frac{\partial^2}{\partial x_l^2} + \frac{\partial^2}{\partial x_m^2} + f\frac{\partial^2}{\partial x_n^2}\right]\sigma_{nn}$$

The purpose of the present section is to review the questions: why is it there and what is its magnitude? The most important thing to grasp, however, is that the term it enters is a very small term; for practical purposes, uncertainty about where f lies between 1 and 0 is of no consequence at all compared with other uncertainties.

We review first the reason for inserting the factor f. The theme of the book is deformation; also change of chemistry is more closely related to creep or nonrecoverable deformation than to elastic or recoverable effects; hence we focus on creep. Atoms have no sense of geography and hence any material that creeps must also be able to self-diffuse: $\partial^2\sigma_{nn}/\partial(R\theta)^2$ and $\partial^2\sigma_{nn}/\partial x_l^2$ must have similar effects. These are the ideas embodied in Figures 11.5 and 11.6. For any direction $\mathbf{n}$, change of σ_{nn} along x_l or along x_m is easy to imagine and to relate to the material's behavior, so that the first four right-hand terms in the equation arise naturally. But as Figure 11.9 shows, change of σ_{nn} along x_n is a fundamentally different phenomenon. In the conditions of Figure 20.4a, what can be said about the tendency of the nonuniform compression to drive material from the central site outward?

It seems reasonable to consider two extremes. First, we ask: can a material be so gaslike, so isotropic in all its behaviors, that the difference $\sigma_{nn}^B - \sigma_{nn}^C$ drives diffusion of atoms from B to C to exactly the same extent as would an equivalent difference $\sigma_{mm}^B - \sigma_{mm}^C$? At this extreme, we admit that there is a geometrical difference between the two parts of Figure 11.9 but postulate that no difference in behavior follows; in this case, $f = 1$.

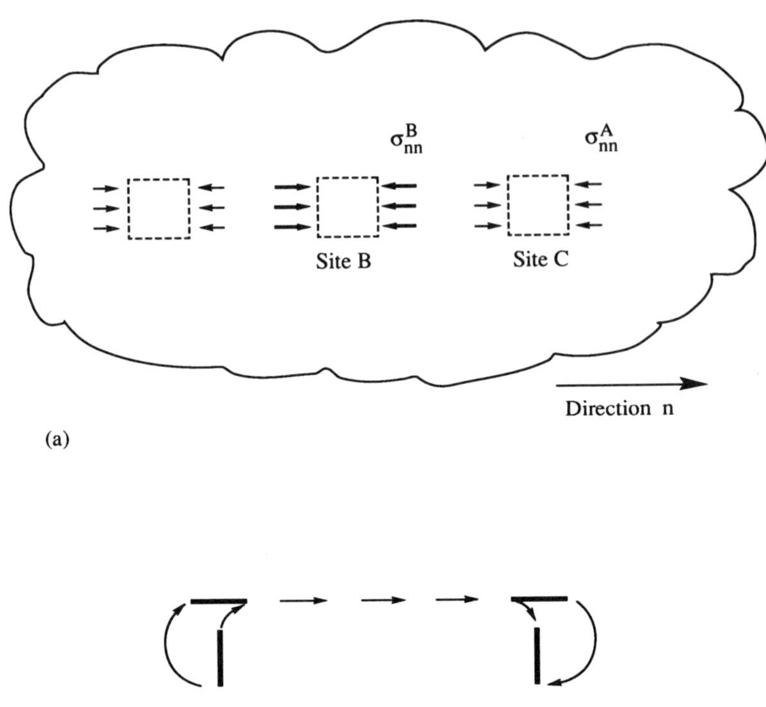

Figure 20.4 (a) Three sites within a stressed continuum where a tendency exists for material to travel outward from site B on account of nonuniform stress. (b) Migration of a wafer from site B to site C could occur by rotation, translation, and a second rotation.

At the other extreme, we ask: can a material be so much a continuum, can planes in the material be so indestructible, that the only response in the conditions of Figure 20.4a is as shown in Figure 20.4b? Without doubt, a difference $\sigma^{A}_{nn} - \sigma^{B}_{nn}$ must entail three separate differences $\sigma^{A}_{nn} - \sigma^{A}_{mm}$, $\sigma^{A}_{mm} - \sigma^{B}_{mm}$, and $\sigma^{B}_{mm} - \sigma^{B}_{nn}$, and these three operating simultaneously would transfer material from site B to site C. But the processes here are all covered by the first four right-hand terms of eqn. (12.7); if the difference $\sigma^{B}_{nn} - \sigma^{C}_{nn}$ was effective only in this manner, we would write $f = 0$.

It seems reasonable to suppose that any real material, whether crystal, glass, or molecular fluid, lies somewhere between these two extremes. The present text concerns materials toward the continuum end rather than the gaslike end, that is, toward the extreme where $f = 0$. In those parts where the material has only one component there is no *need* to imagine gaslike attributes. Also in those parts where a two-component material is treated as a substrate that encompasses an exchange-process, again the substrate needs no gaslike attributes; it is only the exchangers or the imaginary additive that is gaslike. Hence to use the approximation $f = 0$ does not conflict with the other ideas being used. On the other hand, the idea that f, though close to zero, is not exactly zero reflects the fact that no real material is ideally continuous; the gradation of real states from solids to gases seems to require that f be written in.

Fortunately, as discussed on page 125, there seem to be good reasons for expecting $\partial^2\sigma_{nn}/\partial x_n$ to be small compared with $\partial^2\sigma_{nn}/\partial x_m$ in all situations to which this book refers. Hence as long as f is close to zero, we do not need to face the difficulty of establishing exactly what its magnitude is.

Anisotropic materials

The assumption is made throughout the preceding chapters that the materials in view are intrinsically isotropic. Even in Chapter 19 where microstructure is discussed and the suggestion is made that a material's composition might be effectively anisotropic, this is a consequence of the stress state and not a property inherent in the material. But of course a real material may be anisotropic.

For an extreme example, consider the mica *muscovite* (used in manufacturing as easily worked, transparent, heat-resistant sheets). It is essentially an aluminosilicate but larger atoms such as iron or magnesium can substitute for aluminum to a limited extent. Upon substitution, the unit cell dimensions change, two becoming larger; but in one direction the dimension gets smaller as the larger atom enters (Guidotti et al., 1992).

Such a material prompts reconsideration of the ideas embodied in Figure 19.1, especially the idea of treating a material as substrate plus additive, or a gas dissolved in a continuum, with material parameters K^* and K^{a}. The approach is powerful; in Chapters 15 and 16, it facilitates progress; it allows us to suppose that nonhydrostatic stress affects only the substrate, which is chemically uniform, and chemical change is affected only by the mean stress at a point, which is direction-invariant. This separates chemical effects from nonhydrostatic effects; boxes (i) and (ii) in the lower part of Figure 19.1 become empty. But one cannot believe the mica substitution just described is affected only by the mean stress; it must also be affected by the orientation of, for example, the maximum compression, and by the magnitude of the difference $\sigma_{\mathrm{max}} - \sigma_{\mathrm{mean}}$.

The general conclusion is as reached before: if one constructs a body of theory using the separation of chemistry from nonhydrostatic effects, one makes rapid progress over a limited field, but the theory cannot be complete. Nor is it a good basis for further extension; it has more the character of a dead end or trap. A more powerful approach is to assume that in general a chemical component responds to all parts of a nonhydrostatic state of stress. Once this idea is accepted as foundation, the simpler special cases where a component responds only to mean stress can be enjoyed without being the whole of one's universe or roaming-space.

APPENDIX 20A: UNSTEADY BEHAVIOR,
A NUMERICAL EXAMPLE

The purpose of this appendix is to show the means by which curves as in Figure 20.1d could be calculated if a stiffness were known for each of the elements R, S, T, and U (springs and dashpots) in Figure 20.1b.

Let the lengths of the elements be r, s, t, and u; then $r + t = u + s$. Let the lengths of R and S when carrying no load be r_0 and s_0. Let the forces in the strings be p and q newtons; then if a load is imposed as in Figure 20.1c, $p + q = 20$. Also, if a dot is used to show rate of change with time, $\dot{p} = -\dot{q}$. Let the yield factors or inverse stiffness properties of the elements be $4j$ and j in the springs, S being stiffer, and k and $3k$ in the dashpots, T being stiffer. Then

$$s - s_0 = jq \qquad r - r_0 = 4jp$$

$$\dot{s} = j\dot{q} \qquad \dot{r} = 4j\dot{p}$$

$$\dot{t} = kp \qquad \dot{u} = 3kq$$

If the bar remains horizontal $\dot{r} + \dot{t} = \dot{s} + \dot{u}$, or

$$4j\dot{p} + kp = j\dot{q} + 3kq$$

Expressing p in terms of q gives

$$-4j\dot{q} + k(20 - q) = j\dot{q} + 3kq$$

or

$$-5j\dot{q} + 20k - 4kq = 0$$

We need to solve this equation for q.

As discussed in the main text, the magnitude of q at the first moment of loading is 16 newtons, diminishing toward 5 newtons as its ultimate or steady-state magnitude. If we use x for the amount of time since the moment of loading, a suitable expression for q is $(5 + 11e^{-x/A})$, where the characteristic time A has yet to be found. With this expression, $\dot{q} = (-11/A)e^{-x/A}$ and the above equation becomes

$$\frac{55j}{A}e^{-x/A} - 44ke^{-x/A} = 0 \qquad A = \frac{5j}{4k}$$

As an example, let $j = 0.0020$ m/N and $k = 0.0003$ m/N-sec. Then $A = 8.33$ sec and the ultimate descent rate of the bar is 0.0045 m/sec.

Other details are as follows:

1. At the moment of first loading, when $p = 4$ N and $q = 16$ N, the rates of change of p and q are 1.32 N/sec. Rates of elongation of the elements are

$$\dot{r} = 0.01056 \qquad \dot{s} = -0.00264$$
$$\dot{t} = 0.00120 \qquad \dot{u} = 0.01440 \qquad \text{m/sec}$$

The weak elements, R and U, are elongating rapidly; the overall descent rate, 0.01176 m/sec, is about $2\frac{1}{2}$ times the ultimate descent rate.

2. After 25 sec, when $x = 3A$, $e^{-x/A} = 0.05$. At this moment, $q = 5.55$ N and $p = 14.45$ N; that is to say, p and q have already made 95% of the change to their steady-state magnitudes.

3. The moment when forces p and q are equal, at 10 N each, comes when $e^{-x/A} = 5/11$, when $x = 6.6$ sec.

REFERENCES AND NOTES

Some explanatory comments follow the alphabetical list.

Biot, M. A., 1941. General theory of three-dimensional consolidation. *Journal of Applied Physics* **12**, 155–164.

Biot, M. A., 1984. New variational-Lagrangian irreversible thermodynamics with application to viscous flow, reaction-diffusion, and solid mechanics. *Advances in Applied Mechanics* **24**, 1–91.

Bowen, R. M., 1967. Toward a thermodynamics and mechanics of mixtures. *Archive for Rational Mechanics and Analysis* **24**, 370–403.

Bowen, R. M., 1976. Theory of mixtures, in *Continuum Physics*, vol. 3, ed. A. C. Eringen. New York: Academic Press, pp. 1–127.

de Groot, S. R., 1951. *Thermodynamics of Irreversible Processes*. New York: Interscience.

Einstein, A., 1905. Uber die von der molekularkinetischen Theorie der Wärme geforderte Bewegung von in ruhenden Flussigkeiten suspendierten Teilchen. *Annalen der Physik* **17**, 549–560.

Fletcher, R. C., 1982. Coupling of diffusional mass transport and deformation in a tight rock. *Tectonophysics* **83**, 275–291.

Freer, R., 1981. Diffusion in silicate minerals and glasses: a data digest and guide to the literature. *Contributions to Mineralogy and Petrology* **76**, 440–454.

Gibbs, J. W., 1878. On the equilibrium of heterogeneous substances. *Transactions of the Connecticut Academy* **3**, 343–524.

Glasstone, S., Laidler, K. J., and Eyring, H., 1941. *Theory of Rate Processes*. New York: McGraw-Hill, pp. 516–521.

Guidotti, C. V., Mazzoli, C., Sassi, F. P., and Blencoe, J. G., 1992. Compositional controls on the cell dimensions of $2M_1$ muscovite and paragonite. *European Journal of Mineralogy* **4**, 283–297.

Hanley, H. J. M. (ed.), 1969. *Transport Phenomena in Fluids*. New York: Marcel Dekker.

Ozawa, K., 1989. Stress-induced Al–Cr zoning of spinel in deformed peridotites. *Nature* **338**, 141–144.

Ramberg, H., 1959. The Gibbs' free energy of crystals under anisotropic stress, a possible cause for preferred mineral orientation. *Anais da Escola de Minas de Ouro Preto* **32**, 1–12.

Ramberg, H., 1963. Chemical thermodynamics in mineral studies. *Physics and Chemistry of the Earth* **5**, 225–252.

Sane, A. Y., and Cooper, A. R., 1987. Stress buildup and relaxation during ion exchange strengthening of glass. *Journal of the American Ceramic Society* **70**, 86–89.

Stephenson, G. B., 1988. Deformation during interdiffusion. *Acta Metallurgica et Materialia* **36**, 2663–2683.

Terzaghi, K., 1925. Principles of soil mechanics IV settlement and consolidation of clay. *Engineering News-Record* **95**, 874–878.
Vaughan, P. J., Green, H. W., and Coe, R. S., 1984. Anisotropic growth in the olivine-spinel transformation of Mg_2GeO_4 under nonhydrostatic stress. *Tectonophysics* **108**, 299–322.
Weber, N., and Goldstein, M., 1964. Stress-induced migration and partial molar volume of sodium ions in glass. *Journal of Chemical Physics* **41**, 2898–2901.
Woods, L. C., 1975. *The Thermodynamics of Fluid Systems*. Oxford: Clarendon Press.

The first point to explain is the shortness of the list. The subject-matter of the book ranges over chemistry, mechanics, and thermodynamics, and the topic is not well served by any compact and accessible literature; the range of journals and books through which relevant paragraphs are dispersed is exceptionally wide. This is not a fit place to attempt a compilation of this scattered literature; the list above is limited to just three categories of item, as follows.

1. Items on what might be seen as the main stream of ideas leading to the present position, namely those by Terzaghi (1925), Biot (1941), Ramberg (1959, 1963), Bowen (1967), Fletcher (1982), Stephenson (1988).
2. Items intended to be comprehensive and at the same time mathematically rigorous. Examples are the cited works by Biot (1984), Bowen (1976), Hanley (1969), Stephenson (1988), Woods (1975).
3. Items included in support of specific sentences or paragraphs: de Groot (1951), Einstein (1905), Freer (1981), Glasstone et al. (1941), Guidotti et al. (1992), Ozawa (1989), Sane and Cooper (1987), Vaughan et al. (1984).

Reasons for the list being short are, first, the above policy; second, the fact that I do not know the literature well; third, the exclusion of all texts, however high their quality, that treat the mechanisms on atomic scale by which materials move to and fro. As long as the task is to focus on the driving gradients, the fluxes, the constraints, and the interactions, questions about atomic scale mechanisms are both unnecessary and distracting. Papers about mechanisms form, of course, a distinguished literature in their own right, but it is separate from the topics currently in view.

Regarding list 2, readers are entitled to information on the following points: to what extent do the authors all present the same view, and to what extent is the content of the present book compatible with works in that list? My impression is that the reader will find discrepancies all along; what one author writes is different from what another writes. Then the question arises: are the various statements compatible, being different in emphasis or point of view but not in direct conflict so that we can believe them all, or are they incompatible so that if we believe one it is unreasonable at the same time to believe others? Again my impression is that when this question is asked, noone is wholly sure of the answer; there is not yet a basic approach or group of fundamentals on which people are agreed. Readers thus have the privilege, obligation, or pain of having to decide for themselves.

INDEX